Phospholipid
Metabolism
in
Cellular Signaling

José M. Mato, Ph.D.
Instituto de Investigaciones Biomédicas
Consejo Superior de Investigaciones Cientificas
Madrid, Spain

CRC Press
Taylor & Francis Group
Boca Raton London New York

CRC Press is an imprint of the
Taylor & Francis Group, an **informa** business

First published 1990 by CRC Press
Taylor & Francis Group
6000 Broken Sound Parkway NW, Suite 300
Boca Raton, FL 33487-2742

Reissued 2018 by CRC Press

Library of Congress Cataloging-in-Publication Data

Mato, José M., 1949-
    Phospholipid metabolism in cellular signaling/José M. Mato.
        p.   cm.
    Includes bibliographical references.
    Includes index.
    ISBN 0-8493-5978-3
    1. Membrane lipids--Metabolism.   2. Cellular signal transduction.
    3. Cell interaction.   I. Title.
    [DNLM: 1. Cell Communication.   2. Cell Membrane.
    3. Phospholipids--metabolism.   4. Signal Transduction.   QU 93
M433pl
QP752.M45M37   1990
574.87--dc20
DNLM/DLC
for Library of Congress   90-2495

A Library of Congress record exists under LC control number: 90002495

ISBN 13: 978-1-315-89642-7 (hbk)
ISBN 13: 978-1-351-07552-7 (ebk)

Visit the Taylor & Francis Web site at http://www.taylorandfrancis.com and the
CRC Press Web site at http://www.crcpress.com

PREFACE

Over the past few years, the importance of phospholipid metabolism in biological systems has become increasingly recognized, and considerable advances have been made in our understanding of the role of phospholipids in transmembrane signaling. This volume attempts to give a comprehensive and critical view of the more recent advances made in certain major areas of research in the field of phospholipid metabolism in cellular signaling. The volume begins with chapters dedicated to the lipid composition of cellular membranes and to their organization in biological systems (asymmetry in membrane phospholipids and the role of phosphatidylinositol in protein anchoring). The more dynamic aspects of phospholipid metabolism and its regulation by a great variety of hormones and other extracellular stimuli is discussed later (regulation of the metabolism of phosphatidic acid, phosphatidylcholine, and phosphatidylethanolamine). Some major aspects of the breakdown of inositol phosphatides and inositol phosphate metabolism are discussed in Chapter 7, and the role in insulin action of newly discovered glycosyl-phosphatidylinositols is reviewed in Chapter 8. The final two chapters cover two biologically active phospholipids: the sphingolipids and ether-linked glycerophospholipids (platelet activating factor).

José M. Mato

THE AUTHOR

José M. Mato, Ph.D., is Director and Research Professor at the Instituto de Investigaciones Biomédicas of the Consejo Superior de Investigaciones Científicas, Madrid, Spain.

Dr. Mato graduated in 1972 at the University Complutense of Madrid, Madrid, receiving the A.B. degree in biochemistry, and he obtained his Ph.D. degree (summa cum laude) in 1976 from the University of Leiden, Leiden, The Netherlands. He served as an Assistant Professor at the University of Leiden from 1976 to 1978, as an Assistant Professor at the Fundación Jiménez Díaz (Madrid) from 1979 to 1986, as Research Associate Professor and Research Professor of the Instituto de Investigaciones Biomédicas at the Consejo Superior de Investigaciones Científicas (Madrid) from 1986 to 1989, and from 1989 to 1990 as Research Director of the Fundacion Jiménez Díaz (Madrid). He has been visiting scientist at the Biozentrum of the University of Basel, Basel, Switzerland; University of North Carolina, Chapel Hill; National Institutes of Health, Bethesda, Maryland; the University of Pennsylvania, Philadelphia; and at the Memorial Sloan Kettering Cancer Research Center, New York. It was in 1990 that he assumed his present position.

Dr. Mato is a member of the American Association for the Advancement of Science, European Association for the Study of Diabetes, European Society for Clinical Investigation, Spanish Society of Biochemistry, Spanish Society of Endocrinology, Spanish Society for the Study of Liver Diseases, and he is presently serving on the Council of the Fundación Conchita Rábago.

He has received the Kok Prize of the University of Leiden for research in signal transduction, the Novo Prize of the Spanish Society of Endocrinology, and the Morgagni Prize for research in the field of hormonal regulation of phospholipid metabolism. He has been the recipient of many research grants from the Comisión Interministerial de Ciencia y Tecnología, the Fondo de Investigaciones Sanitarias, and from the Program for Research and Development of Spain.

Dr. Mato is the author of more than 100 papers. His current major research interests relate to the hormonal regulation of phospholipid metabolism and to the molecular pathogenesis of the liver in animals and humans.

CONTRIBUTING AUTHOR

Isabel Varela

Isabel Varela, Ph.D., is a member of the Instituto de Investigaciones Biomédicas of the Consejo Superior de Investigaciones Científicas, Madrid, Spain.

Dr. Varela graduated in 1981 at the University Complutense of Madrid, receiving the A.B. degree in biochemistry, and she obtained her Ph.D. degree (summa cum laude) in 1985 from the same University. In 1986 she served as a post-doctoral fellow at the University of Glasgow, Scotland, U.K. and as Research Assistant at the Fundación Jiménez Díaz (Madrid) from 1987 to 1989. She has been visiting scientist at the Institute of Cellular Pathology of the Catholic University of Louvaine, Brussels, Belgium.

Dr. Varela is a member of the Spanish Society of Biochemistry and of The Biomedical Society, U.K. She is author of 29 papers, and her current major research interest relates to the regulation of phospholipid metabolism by insulin and NGF.

ACKNOWLEDGMENTS

The research program of José M. Mato is supported by the Comision Interministerial de Ciencia y Tecnologia, Fondo de Investigaciones Sanitarias, and Europharma. We thank Jorge Amich for the drawings.

Dedicated to Cristina

TABLE OF CONTENTS

Chapter 1

LIPID COMPONENTS OF CELLULAR MEMBRANES

José M. Mato

TABLE OF CONTENTS

I. MEMBRANE COMPONENTS

All cell membranes consists mostly of an association of lipids and proteins. The proportion of both components differs between different membranes and is not even constant in time for a given membrane. Table 1 shows the ratio of protein to lipid for various rat liver membranes.[1] Whereas plasma membranes isolated from rat liver have a protein/lipid ratio of about 1.5, this same ratio in the inner membrane of mitochondria is 3.6. These differences are due to the specific functions that each membrane plays within the cell. As a general rule, membranes which are metabolically more active contain a higher proportion of protein. Certain proteins or lipids are almost exclusively associated with one type of membrane and do not appear in other morphologically different membranes. These specific components are therefore "membrane markers" and can be used to assess the purity of particular membrane fractions. Membrane markers most commonly used possess enzyme activity. Some of the enzyme activities commonly used during purification of rat liver membranes are shown in Table 2. Several lipids are plasma membrane markers. Thus, the phosphoinositides phosphatidylinositol-4-phosphate and phosphatidylinositol-4,5-bisphosphate and the galactosides are almost exclusively localized in the plasma membrane.[2-4] Mitochondrial membranes are characterized by the presence of diphosphatidylglycerol, which is mainly located in the inner membrane of this subcellular fraction. The relative amount of a given marker (protein or lipid) might vary during the life cycle of the cell. These changes are associated with the various functions of a given membrane during the life cycle of the cell.

In addition to the lipid and protein, biological membranes contain carboydrates (up to 10% of their dry weight), water (about 20% of the total mass), and magnesium, calcium, and other ions. Carbohydrates are preferentially found on the cell surface, where they are covalently associated with proteins and lipids.

II. LIPID COMPONENTS OF CELLULAR MEMBRANES

A. GENERAL INFORMATION

The hydrophobic nature of cell membranes is due to the lipid components, which are water insoluble. Four major classes of lipids can be distinguished in eukaryotic cell membranes: glycerophospholipids, sphingolipids, glycosphingolipids, and sterols. Glycerophospholipids and sphingolipids form a group known as phospholipids. Two distinct moieties can be differentiated in these four classes of lipids: a hydrophobic moiety and a polar headgroup of hydrophillic substituent (Figures 1 and 2). The length of this polar headgroup varies between the different lipids from a single hydroxyl group in a molecule of cholesterol to a glycoprotein with an M_r of several thousands in the glycophospholipids that serve as protein anchors to the membrane. Triglycerides, which had not been thought to be membrane components and which do not have a polar headgroup, have also been detected in highly purified plasma membranes. Interestingly, these neutral lipids make up about 6% of the lipid content of plasma membranes from malignant cells.[5]

B. GLYCEROPHOSPHOLIPIDS

Glycerophospholipids are the most ubiquitous and abundant group of membrane lipids.[6-8] They contain as a common structural backbone a diacylglycerol with a phosphate esterified to the primary hydroxyl group of the *sn*-glycerol moiety (Figure 1).[9] With the exception of phosphatidic acid, this phosphate forms a phosphodiester bond with the hydroxyl group of a polar headgroup substituent. The most abundant substituents in eukaryotic cell membranes are choline, ehtanolamine, serine, glycerol, and *myo*-inositol (Figure 1).

Phosphatidylcholine (PC) and phosphatidylethanolamine (PE) are zwitterionic and have no net charge at physiological pH. The headgroup of these phosphoglycerolipids contains

TABLE 1
Protein Lipid Ratio of Rat Liver Membranes

Membrane	Protein/lipid
Plasma membrane	2.0
Endoplasmic reticulum	2.3
Mitochondrial membranes	
Inner membrane	3.6
Outer membrane	1.2
Golgi membranes	2.4

Data from Quinn, P. J., *The Molecular Biology of Cell Membranes*, 1st ed., MacMillan, London, 1976, chap. 1.

TABLE 2
Enzyme Markers of Various Membrane Fractions Isolated from Rat Liver

Enzyme	Membrane
5′-Nucleotidase	Plasma membrane
Glucose 6-phosphate	Endoplasmic reticulum
Succinate reductase	Mitochondria
UDP-galactose:N-acetyl-glucosamine galactosyl transferase	Golgi apparatus
Acid phosphatase	Lysosomal membrane

the negative charge of the phosphate group and the positive charge of the bases. Phosphatidylinositol (PI), phosphatidylglycerol (PG), and phosphatidic acid (PA) carry a net negative charge of -1 at physiological pH, due to the ionization of the phosphate group. Phosphatidylserine (PS) has two types of negative charges, the phosphate and carboxyl groups, and one positively charged amino group. The net charge of this phosphoglycerolipid is -1. Diphosphatidylglycerol, also called cardiolipin, contains a net negative charge of -2 (Figure 2). Phosphatidylinositol exists in various forms with additional phosphates, the most abundant being the phosphatidylinositol-4-phosphate (PtdIns 4-P) and phosphatidylinositol-4,5-bisphosphate (PtdIns 4,5-P$_2$). The headgroups of these phosphoinositides contain net charges of -3 and -5, respectively. A structurally related class of phospholipids are the glycosyl phosphatidylinositols. This group of lipids contains a glucosaminyl oligosaccharide glycosidically bound to the inositol residue. These molecules function as protein anchors to the membrane and also seem to be involved in cellular signaling. Evidently, these differences in charge of the various polar headgroups confer specific structural properties to the various classes of glycerophospholipids which are important in determining the structure of the membranes.

There is a large variation in the fatty acids for a given class of glycerophospholipids.[10] The fatty acids commonly found in membrane lipids are shown in Table 3. In the majority of mammalian membranes, an unsaturated fatty acid is found in position 2 of the glycerol backbone and a saturated fatty acid in position 1. The most abundant saturated fatty acids are palmitic (16:0) and stearic (18:0) acid and the most common unsaturated fatty acids are oleic (18:1), linoleic (18:2), linolenic (18:3), and arachidonic acid (20:4). Exceptions to this rule are the disaturated phosphatidylcholine and phosphatidylglycerol species found in pulmonary surfactant and in the glycosyl phosphatidylinositol that serves as protein anchors to the membrane.

In Figure 1 is represented the structure of a phosphatidylcholine molecule esterified with stearic acid in position 1 and with oleic acid in position 2. The latter has a *cis* double bond between carbons 9 and 10. It is obvious from Figure 2 that an unsaturated fatty acid, such as oleic acid, cannot pack as tightly as a saturated fatty acid, such as stearic acid, does in the membrane. This confers to the molecules of unsaturated fatty acids specific properties

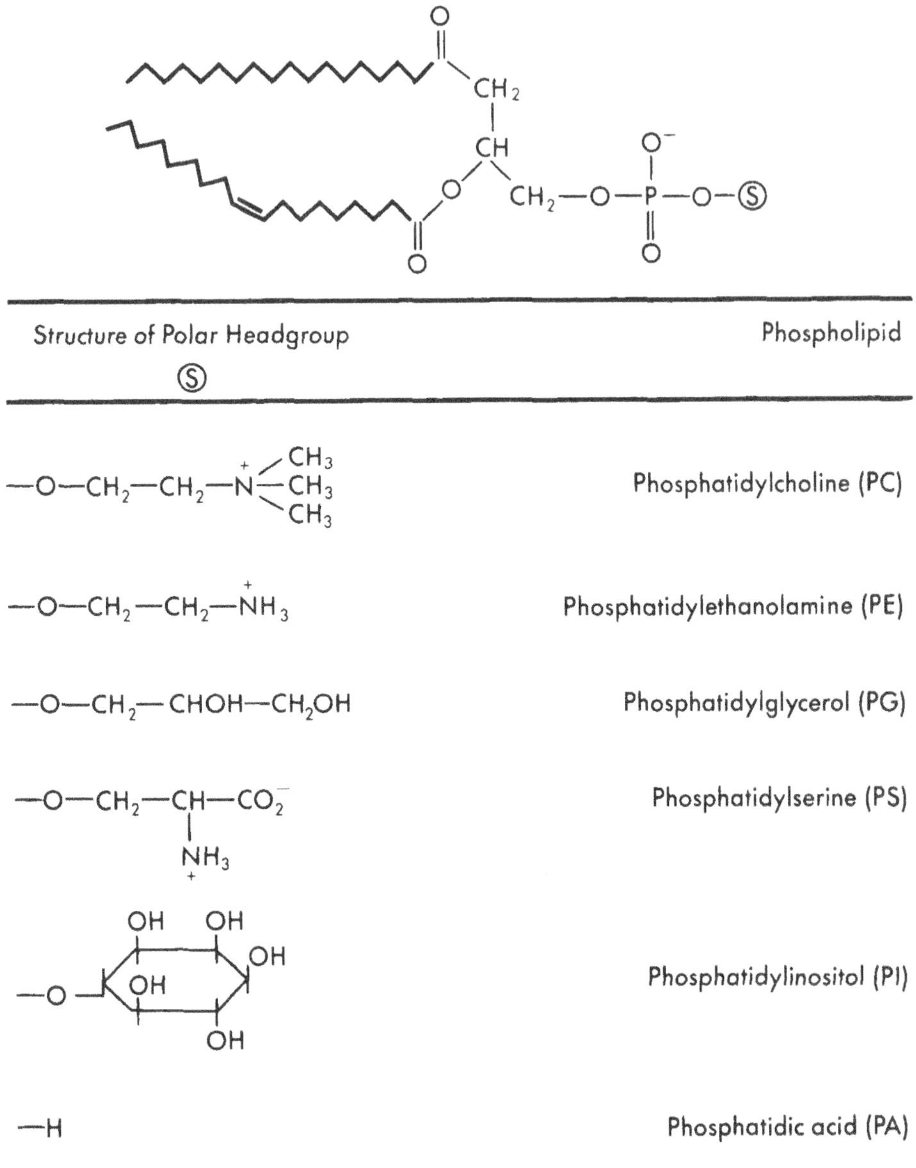

Structure of Polar Headgroup Ⓢ	Phospholipid
—O—CH$_2$—CH$_2$—$\overset{+}{N}\big<\begin{smallmatrix}CH_3\\CH_3\\CH_3\end{smallmatrix}$	Phosphatidylcholine (PC)
—O—CH$_2$—CH$_2$—$\overset{+}{N}H_3$	Phosphatidylethanolamine (PE)
—O—CH$_2$—CHOH—CH$_2$OH	Phosphatidylglycerol (PG)
—O—CH$_2$—CH—CO$_2^-$ │ $\overset{+}{N}H_3$	Phosphatidylserine (PS)
(inositol ring structure)	Phosphatidylinositol (PI)
—H	Phosphatidic acid (PA)

FIGURE 1.   Structure of the most common glycerophospholipids in mammalian biomembranes.

which are important in determining the structure of the biological membrane in which they are immersed.

Another class of glycerophospholipids are the lysophospholipids. These molecules lack a fatty acid in either position 1 or 2 of the glycerol backbone. In biological membranes, these molecules are present in small amounts and are considered as intermediates in phospholipid synthesis or degradation.

A structurally related group of glycerophospholipids are the alkyl ether and the alk-1-enyl ether (also called plasmalogens) phospholipids (Figure 2). These molecules possess a long-chain fatty alcohol in ether linkage to the 1-position of the glycerol moiety of the lipid and might serve to protect animal cells against photosensitized killing.[21,22] A most interesting member of this class of phospholipids, for its biological properties as platelet activator, is the 1-O-octadecyl/hexadecyl-2-acetyl-*sn*-glycero-3-phosphocholine (see Chapter 10). The most common headgroups of these phospholipid subclasses are ethanolamine and choline.

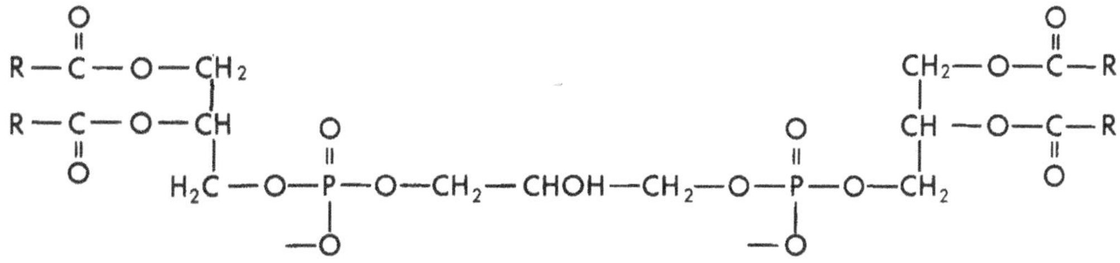

Diphosphatidylglycerol or Cardiolipin

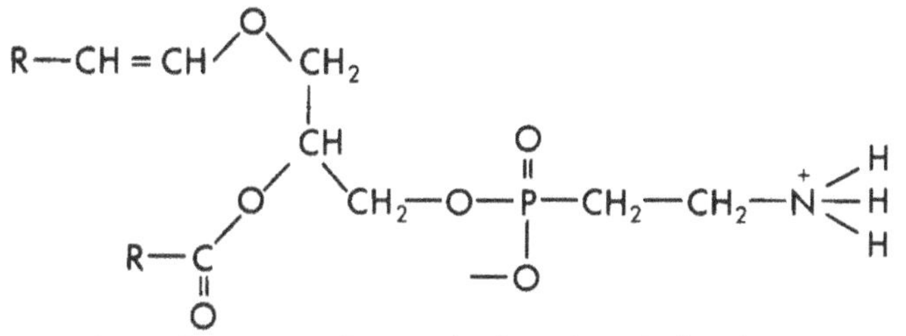

Plasmalogen or Alk-1-enyl-ether phospholipid

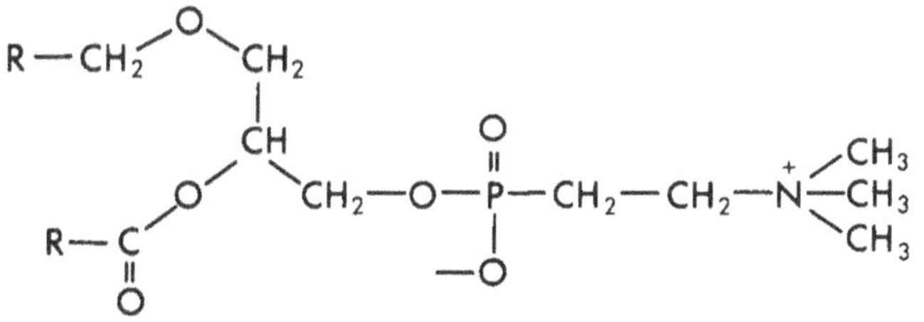

Alkyl-ether phospholipid

FIGURE 2. Structure of diphosphatidylglycerols or cardiolipin, plasmalogens (alk-1-enyl-ether glycerophospholipids) and alkyl-ether glycerophospholipids.

TABLE 3
Common Fatty Acid Moieties Occurring in Glycerophospholipids

Structure	Notation	Trivial name
$CH_3(CH_2)_{14}CO_2H$	16:0	Palmitic acid
$CH_3(CH_2)_{16}CO_2H$	18:0	Stearic acid
$CH_3(CH_2)_5CH=CH(CH_2)_7CO_2H$	16:1	Palmitoleic acid
$CH_3(CH_2)_7CH=CH(CH_2)_7CO_2H$	18:1	Oleic acid
$CH_3CH_2(CH=CHCH_2)_3(CH_2)_6CO_2H$	18:2	Linoleic acid
$CH_3CH_2(CH=CHCH_2)_3(CH_2)_6CO_2H$	18:3	Linolenic acid
$CH_3(CH_2)_4(CH=CHCH_2)_4(CH_2)CO_2H$	20:4	Arachidonic acid

The terms ethanolamine glycerolipids, and choline glycerolipids are used to include the diacyl, alkylacyl, and alk-1-enylacyl forms of these lipids.

C. SPHINGOLIPIDS

This group of compounds was described in 1884 by Thundicum. The simplest class of sphingolipids are sphingomyelins, a major component of animal cell membranes and serum lipoproteins. The structure of sphingomyelin, N-acylsphinogosine-1-phosphorylcholine, or

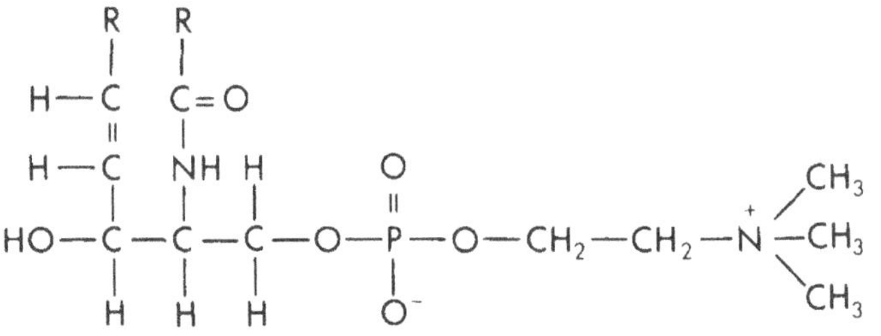

Sphingomyelin

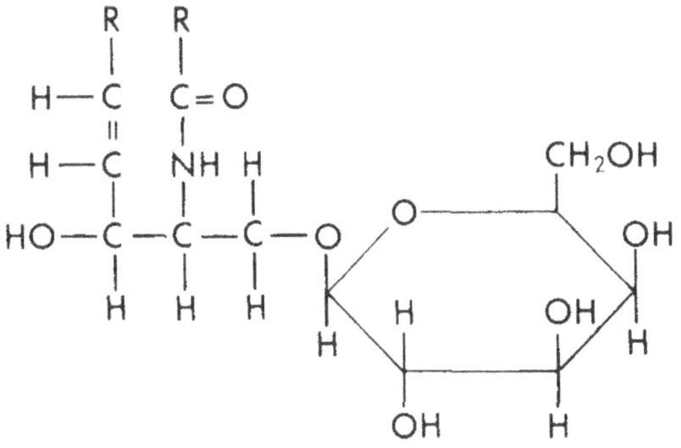

Glycosphingolipids

FIGURE 3.   Structure of sphingomyelins and glycosphingolipids.

ceramide-1-phosphorylcholine, is shown in Figure 3.[11,12] As the phosphatidylcholine molecules, sphingomyelins consist of two hydrophobic groups (the sphingosine group and the acyl group) and a phosphorylcholine group. Sphingomyelins occurring in biological membranes differ in the nature of the sphingosine base and in the acyl group. The most common sphingosine is the 18-carbon aminediol, 1,3-dihydroxy-2-amino-4-octodecene. This molecule has a *trans* double bond between carbons 4 and 5. The dihydro derivative of this base, 1,3-dihydroxy-2-amino-octadecane, is also present in biological membranes in small amounts.[13-15] Phytosphingosine (1,3,4-hydroxy-2-aminooctadecane) has also been detected in bovine kidney.[16] The fatty acids commonly found in sphingomyelin are palmitic (16:0), nervonoyl (24:1, 22:0, and 24:0) acids.[17] In brain, the most common fatty acid found is stearic acid (18:0), nervonoyl (24:1) and (24:0) being less common.[18] An interesting difference in the fatty acid composition of phosphoglycerolipids is the presence of longer fatty acid molecules (22:0, 24:0, and 24:1) in sphingomyelins.

D. GLYCOSPHINGOLIPIDS

Glycosphingolipids are a group of ubiquitous membrane lipids formed from an *N*-acyl sphinogosine, a ceramide, glycosidically bound to a single hexose or a complex oligosaccharide molecule through the hydroxyl group at the C-1 position (Figure 3). The sphingosine and acyl groups form the hydrophobic core of the molecule and the hexoses or the oligosaccharide moiety contributes to the hydrophillic portion. Five subclasses of glycosphingolipids can be differentiated with basically different oligosaccharide moieties. These are named ganglio-, globo-, lacto-, gala-, and mucoglycoceramides. Within each subclass, variations in the oligosaccharide chain give rise to a large variety of glycosphingolipids, particulary in the lacto series. The oligosaccharide structure of several glycosphingolipids

TABLE 4
Basic Structures of the Main Classes of Glycolipids

Name of oligosaccharide	Structure
Gangliotetraose	Galβ1 → 3GalNAcβ1 → 4Galβ1 → 4Glcβ1 → 1Cer ‖ 2αNeuAc 2αNeuAc
Globo-neo-tetraose	GalNAcβ1 → 3Galα1 → 4Galβ1 → 4Glcβ1 → 1Cer
Lactotetraose	Galβ1 → 3GlcNAcβ1 → Galβ1 → 4Glcβ1 → 1Cer
Galabiose	Galα1 → 4Galβ1 → 1Cer
Mucotetraose	Galβ1 → 3Galβ1 → 3Galβ1 → 4Glcβ1 → 1Cer ‖ 1αFuc 1βGal3 ← 1αGalNAc

Note: Cer, ceramide.

is shown in Table 4. All gangliosides have a negative charge at physiological pH due to the presence of sialic acid residues. Detailed descriptions of the glycosphingolipids can be found in several reviews.[19,20]

REFERENCES

1. **Quinn, P. J.**, *The Molecular Biology of Cell Membranes*, 1st ed., MacMillan, London, 1976, chap. 1.
2. **Rawyler, A. J., Roelofsen, B., Wirtz, K. W. A., and Op den Kamp, J. A. F.**, (Poly) phosphoinositide phosphorylation is a marker for plasma membrane in Friend erythroleukaemic cells, *FEBS lett.*, 148, 140, 1982.
3. **Hakomori, S.**, Glycosphingolipids in cellular interaction, differentiation, and oncogenesis, *Ann. Rev. Biochem.*, 50, 733, 1981.
4. **Barbosa, M. L. F. and Pinto da Silva, P.**, Restriction of glycolipids to the outer half of a plasma membrane: concanavalin A labeling of membrane halves in Acanthamoeba castellanii, *Cell*, 33, 959, 1983.
5. **May, G. L., Wright, L. C., Holmes, K. T., Williams, P. G., Smith, I. C. P., Wright, P. E., Fox, R. M., and Mountford, C. D.**, Assignment of methylene proton resonances in NMR spectra of embryonic and transformed cells to plasma membrane triglycerides, *J. Biol. Chem.*, 261, 3048, 1986.
6. **Ansell, G. B. and Hawthorne, J. N.**, Eds., *Phospholipids, Chemistry, Metabolism and Function*, Elsevier/North-Holland, Amsterdam, 1964.
7. **McMurray, W. C. and Magee, W. L.**, Phospholipid metabolism, *Annu. Rev. Biochem.*, 41, 129, 1972.
8. **Rouser, G., Nelson, G. J., Fleischer, S., and Simon, G.**, Lipid composition of animal cell membranes, organelles and organs, in *Biological Membranes, Physical Fact and Function*, Chapman, D. Ed., Academic Press, London, 1968, 5.
9. **Hawthorne, J. N. and Ansell, G. B.**, Eds., *Phospholipids*, Elsevier/North-Holland, Amsterdam, 1982.
10. **Kuksis, A.**, Ed., Fatty acids and glycerides, *Handbook of Lipid Research*, Vol. 1, Plenum Press, New York, 1978.
11. **Pick, L. and Bielschowsky, M.**, Über lipoidzellige splenomegalie (typus Niemann-Pick) und amaurotische idiotie, *Klin. Wochenschr.*, 6, 1631, 1927.
12. **Shapiro, D. and Flowers, H. M.**, Studies on sphingolipids. VII. Synthesis and configuration of natural sphingomyelins, *J. Am. Chem. Soc.*, 84, 1047, 1962.
13. **Sweeley, E. L. and Moscatelli, E. A.**, Qualitative microanalysis and estimation of sphingolipid bases, *J. Lipid Res.*, 1, 40, 1959.
14. **Hirvisalo, E. L. and Renkonen, O.**, Composition of human serum sphingomyelins, *J. Lipid Res.*, 11, 54, 1970.
15. **Samuelsson, B. and Samuelsson, K.**, Separation and identification of ceramides derived from human plasma sphingomyelins, *J. Lipid Res.*, 10, 47, 1969.
16. **Karlsson, K. A. and Steen, G. O.**, Studies on sphingosines. XIII. The existence of phytosphingosine in bovine kidney sphingomyelins, *Biochim. Biophys. Acta*, 152, 789, 1968.
17. **Svennerholm, E., Stallberg-Stenhagen, S., and Svennerholm, L.**, Fatty acid composition of sphingomyelins in blood, spleen, placenta, liver, lung and kidney, *Biochim. Biophys. Acta*, 125, 60, 1966.

18. **O'Brien, J. S. and Rouser, G.**, The fatty acid composition of brain sphingolipids: sphingomyelin, ceramide, cerebroside, and cerebroside sulfate, *J. Lipid Res.*, 5, 339, 1964.
19. **Hakomori, S.**, Glycosphinogolipids in cellular interaction differentiation and oncogenesis, in *Handbook of Lipid Research: Sphingolipid Biochemistry*, Vol. 3, Hannahan, D. J., Ed., Academic Press, New York, 1983, 327.
20. **Wiegandt, H.** Gangliosides, in *Glycosphingolipids*, Wiegandt, H., Ed., Elsevier, Amsterdam, 1985, 199.
21. **Morand, O. H., Zoeller, R. A., and Raetz, C. R. H.**, Disappearance of plasmalogens from membranes of animal cells subjected to photosensitized oxidation, *J. Biol. Chem.*, 263, 11597, 1988.
22. **Zoeller, R. A., Morand, O. H., and Raetz, C. R. H.**, A possible role for plasmalogens in protecting animal cells against photosensitized killing, *J. Biol. Chem.*, 263, 11590, 1988.

Chapter 2

PHOSPHOLIPID COMPOSITION OF CELLULAR MEMBRANES

José M. Mato

TABLE OF CONTENTS

I. GENERAL INFORMATION

As mentioned in Chapter 1, biological membranes contain a large variety of lipid groups (glycerophospholipids, sphingolipids, glycolipids, and others), variants within each group (glycophospholipid classes: phosphatidylcholine, phosphatidylethanolamine, and others), and different molecular species within each class (phosphatidylcholine molecules with different fatty acid composition). The development of this large variety of lipid molecules must have resulted in evolutionary advantages for the biological membranes to perform their functions. This suggests the existence of specific roles for certain lipid molecules (e.g., phosphatidylinositol-4,5-bisphosphate, found almost exclusively in the plasma membrane and the precursor of inositol-1,4,5-trisphosphate, a modulator of intracellular calcium levels, and diacylglycerol, a modulator of protein kinase C), or for a given lipid composition (e.g., the lipid composition of mitochondria does not vary significantly between tissues of the same species). However, there are also large variations in lipid composition between membranes from different tissues of one given species that perform a similar function (e.g., the ratio of sphingomyelin to phosphatidylcholine of the endoplasmic reticulum from bovine kidney, heart, and liver varies largely). This indicates that although certain lipids or lipid mixtures might have specific functions, a similar function can be performed by a variety of specific lipid compositions. The mechanisms by which a specific lipid composition is achieved and maintained in a particular biological membrane is one of the most challenging questions in molecular membrane biology.

II. PHOSPHOLIPID COMPOSITION OF BIOLOGICAL MEMBRANES

A. PHOSPHOLIPID COMPOSITION OF WHOLE ORGANS

Rat liver contains 32 to 40 μmol phospholipid per gram of wet tissue.[1-3] The phospholipid composition of total rat liver is shown in Table 1.[1,3-6] Phosphatidylcholine, phosphatidylethanolamine, phosphatidylinositol, and sphingomyelin account for about 90% of the total rat liver phospholipids. Diphosphatidylglycerol, phosphatidylserine, and phosphatidic acid are present in concentrations ranging from 5 to 1%. Other phospholipids (i.e., phosphatidylglycerol, phosphatidylinositol-4-monophosphate, and phosphatidylinositol-4,5-bisphosphate) account for less than 1% of the total cell phospholipids. In other words, there is around 100 μg of these less abundant phospholipids per gram of wet rat liver.

The phospholipid composition of other rat organs does not differ drastically from that of the liver. Table 1 compares the phospholipid composition of rat liver with that of rat testes[5] and skeletal muscle.[6] As can be seen, phosphatidylcholine, phosphatidylethanolamine, phosphatidylinositol, and sphingomyelin also account for about 90% of the total phospholipids in both rat testes and skeletal muscle.

B. PHOSPHOLIPIDS IN SUBCELLULAR MEMBRANES

The most abundant subcellular membrane in rat liver is the endoplasmic reticulum, which accounts for about 50% of the membrane mass; mitochondria and Golgi membranes account for about 30 and 6%, respectively, and plasma membranes and lysosomal membranes for about 5 and 1%, respectively.[7,8] Data on the phospholipid composition of these subcellular fractions are summarized in Table 2.[1,9,10] Each subcellular fraction has a specific lipid composition. Rough endoplasmic reticulum and nuclear membranes contain about 60% of the total phosphatidylcholine, while plasma membranes and secondary lysosomes contain around 37%. The opposite is true for sphingomyelin: secondary lysosomes and plasma membranes contain 32.9 and 18.1%, of the total sphingomyelin, respectively, while the endoplasmic reticulum and nuclear membranes contain about 3.5%, and in cultured human

TABLE 1
PHOSPHOLIPID COMPOSITION OF SEVERAL RAT
TISSUES

Phospholipid classes	Percent of total phospholipid		
	Rat liver[a]	Rat testes[b]	Rat skeletal muscle[c]
PC	49.4	44.6	51.1
PE	23.8	25.8	22.2
SL	5.7	6.3	2.7
PI	8.1	nr	8.9
PS	3.0	5.6	3.7
PG	0.8	nr	0.9
DPG	5.1	nr	1.4
PA	1.5	nr	0.8

Note: PC and PE includes the acyl, alkyl, and alkenyl subclasses. PI includes the mono-, di-, and triphosphoinositides. PC, phosphatidylcholine; PE, phosphatidylethanolamine; SL, sphingomyelin; PI, phosphatidylinositol; PS, phosphatidylserine; PG, phosphatidylglycerol; DPG, diphosphatidylglycerol; PA, phosphatidic acid.

[a]   Data from References 1, 3, and 4.
[b]   Data from Reference 5.
[c]   Data from Reference 6.

fibroblasts, plasma membranes contain 90% of the total sphingomyelin.[31] The content of phosphatidylethanolamine is about 20 to 25% in all subcellular fractions. The amount of phosphatidylinositol is also relatively constant, with a value for all subcellular fractions of about 7 to 9%. Mitochondria have a characteristically high diphosphatidylglycerol content (15.8%). Plasma membranes have a relatively high content in phosphatidic acid (4.4%) and Golgi membranes have a high proportion of lysophospholipids (about 12%), which is compatible with this molecule being formed during lipid degradation. The phospholipid composition of nuclear membranes is very similar to that of the endoplasmic reticulum and, in agreement with their endocytic origin, secondary lysosomes have a phospholipid composition similar to that of plasma membranes.

Table 3 summarizes the data on the phospholipid composition of the inner and outer mitochondria membranes.[10] These two membranes are remarkably different not only in their lipid composition, but also in their protein to lipid ratio (see Table 1, Chapter 1) and in the cholesterol to protein ratio in the inner membrane (0.02) and in the outer membrane (0.04).[10] Both membranes have a high content of diphosphatidylglycerol, which is, however, higher in the inner membrane. Both membranes have a similar content of phosphatidylcholine and phosphatidylethanolamine, but the outer membrane resembles more the endoplasmic reticulum, with a higher content of sphingomyelin, phosphatidylinositol, and phosphatidylserine.

In addition to this specific localization of certain phospholipids in a given membrane fraction, it is important to note, for example, that plasma membranes contain most of the cellular glycolipids (these lipids are not commonly found in mitochondria, endoplasmic reticulum, or nuclear membranes) and cholesterol.[10,31] The cholesterol to phospholipid ratio of rat liver plasma membrane is 0.6, whereas that in lysosomes, Golgi, endoplasmic reticulum, and mitochondria is 0.49, 0.31, 0.12, and 0.1, respectively.

As mentioned in Chapter 1, choline and ethanolamine glycophospholipids are formed by three subclasses: diacyl, alkylacyl, and alk-1-enylacyl glycerophospholipids. Ether-linked lipids occur throughout the animal kingdom and are also found as minor membrane lipids of higher plants. As a rule, alkyl groups are mostly associated with choline glycerophospholipids and alk-1-enyl groups are almost exclusively associated with ethanolamine gly-

TABLE 2
PHOSPHOLIPID COMPOSITION OF RAT LIVER CELL MEMBRANE

Phospholipid	Percent of total phospholipid					
	Rough endoplasmic reticulum	Nuclei	Mitochondria	Golgi	Plasma membrane	Secondary lysosomes
PC	60.5	61.4	40.9	45.6	37.1	37.7
PE	20.9	22.7	33.4	17.8	18.6	19.2
SL	3.8	3.2	2.4	12.5	18.1	32.9
PI	9.0	8.6	6.5	8.6	7.0	7.4[a]
PS	3.3	3.6	0.9	4.2	6.2	
PG	nd	nd	2.3	nd	4.8	nd
DPG	1.2	nd	15.8	nd	traces	6.8[b]
PA	nd	<1	1	nd	4.4	
LPC	3.1	1.5	1.4	5.9	1.9	0
LPE	0	0	nd	6.3	nd	nd

Note: All values are expressed as μmoles percent of total phospholipid. PC, phosphatidylcholine; PE, phosphatidylethanolamine; SL, sphingomyelin; PI, phosphatidylinositol; PS, phosphatidylserine; PG, phosphatidylglycerol; DPG, diphosphatidylglycerol; PA, phosphatidic acid; LPC, lysophosphatidylcholine; LPE, lysophosphatidylethanolamine. nd, Not detectable.

[a] Value for PI + PS.
[b] Value for DPG + PA.

Data from References 1, 9, and 10.

13

TABLE 3
PHOSPHOLIPID COMPOSITION OF RAT LIVER MITOCHONDRIA

	Percent of total phospholipid (mitochondria)	
Phospholipid	Inner membrane	Outer membrane
PC	45.4	49.7
PE	25.3	23.2
SL	2.5	5.0
PI	5.9	12.6
PS	0.9	2.2
PG	2.1	2.5
DPG	17.4	3.4
PA	0.7	1.3

Note: PC, phosphatidylcholine; PE, phosphatidylethanolamine; SL, sphingomyelin; PI, phosphatidylinositol; PS, phosphatidylserine; PG, phosphatidylglycerol; DPG, diphosphatidylglycerol; PA, phosphatidic acid.

Data from Reference 10.

TABLE 4
AMOUNT OF DIACYL, ALKYLACYL, AND ALK-1-ENYLACYL SUBCLASSES PRESENT IN THE CHOLINE AND ETHANOLAMINE GLYCEROPHOSPHOLIPIDS

	Percent of total subclass			
Subclass	Rat erythrocytes	Human endothelial cells	Beef brain	Bovine heart
diacyl-GPC	97.6	88.6	nr	58.1
alkylacyl-GPC	0.9	4.6	nr	nr
alkenylacyl-GPC	0.5	6.8	nr	41.9
diacyl-GPE	39.6	52.6	29.8	51.2
alkylacyl-GPE	6.8	4.9	3.5	nr
alkylacyl-GPE	53.6	42.5	66.7	48.8

Note: GPC, glycerophosphocholine; GPE, glycerophosphoethanolamine. nr, Not reported.

Data from References 11 through 14.

cerophospholipids. An exception to this rule is heart tissue, where a large proportion of the alk-1-enyl groups are associated to choline glycerophospholipids.

Table 4 shows the relative amount of diacyl, alkylacyl, and alk-1-enylacyl subclasses present in the choline and ethanolamine glycerophospholipids of rat erythrocytes, human endothelial cells, beef brain, and bovine heart.[11-14] It is important to note that in these four different tissues, the ether-linked lipids account for about 50% of the ethanolamine glycerophospholipids. The relative amount of ether-linked lipids in choline glycerophospholipids varies from 1.4% in rat erythrocytes to 41.9% in bovine heart.

C. FATTY ACID COMPOSITION OF GLYCEROPHOSPHOLIPID CLASSES

The fatty acid composition of the major glycerophospholipid classes of rat liver is shown in Table 5.[15-22,30] Each phospholipid class has a specific pattern of fatty acids. The most

TABLE 5
FATTY ACID COMPOSITION OF RAT LIVER
GLYCEROPHOSPHOLIPIDS

Fatty acid	Percent of total						
	PC	PE	PI	PS	PG	DPG	PA
16:0	23.6	22.9	4.3	22.8	12.0	4.4	28.6
16:1	0.5	0.4	0.2	2.2	1.8	2.6	0.8
18:0	25.5	27.6	27.4	15.3	14.1	1.6	17.0
18:1	7.3	6.6	17.0	15.0	20.9	14.9	20.7
18:2	14.1	9.8	7.8	30.9	19.7	76.5	20.8
20:4	22.6	22.8	36.3	7.6	29.9	nd	11.1
22:6	4.6	9.9	0.9	2.0	nd	nd	0.6

Note: PC, phosphatidylcholine; PE, phosphatidylethanolamine; PI, phosphatidylinositol; PS, phosphatidylserine; PG, phosphatidylglycerol; DPG, diphosphatidylglycerol; PA, phosphatidic acid.

Data from References 15 through 22.

abundant saturated fatty acids in phosphatidylcholine and phosphatidylethanolamine are
palmitic (16:0) and stearic (18:0) acid, which account for about 50% of the total fatty acids
of these two glycerophospholipids. The most abundant unsaturated fatty acid in phosphatidycholine and phosphatidyethanolamine is arachidonic acid (20:4) (about 23%). These two
glycerophospholipids differ by a relative amount of 22:6, which is two times more abundant
in phosphatidylethanolamine (9.9%) than in phosphatidylcholine (4.6%). Phosphatidylinositol contains the highest proportion of 20:4 (36.3%) and a low percent of 16:0. Phosphatidylserine has a low percent of 20:4, but a high proportion of 18:2. The amount of 18:2 is
highest in diphosphatidylglycerol, where about 75% of the fatty acids are of this type. Since
diphosphatidylglycerol is almost exclusively located in the inner mitochondrial membrane
(Table 3), it is obvious that this particular membrane has a very specific lipid composition,
with respect to both the polar headgroup and the fatty acid composition of its lipids. This
might confer to this membrane specific properties important to the performance of its functions.

D. FATTY ACID COMPOSITION OF GLYCEROPHOSPHOLIPID SUBCLASSES

In addition to these differences in fatty acid composition between the various glycophospholipid classes, there are also marked differences among the different subclasses for
a given glycophospholipid (i.e., diacyl, alkylacyl, and alk-1-enylacyl glycerophosphocholine). Thus, in human neutrophils, where the predominant pools of arachidonic acid are
ethanolamine (68%), choline (19%), and inositol (12%) glycerophospholipids, alkylacyl
glycerophosphocholine contained 66% of the arachidonic acid, and in the case of ethanolamine glycerophospholipids, 71% of the arachidonic acid was in the form of alk-1-enylacyl
glycerophosphoethanolamine.[23] Interestingly, alk-1-enylacyl glycerophosphoethanolamine is
the major storage place for arachidonic acid in rabbit vascular smooth cells, and this pool
is rapidly hydrolyzed during cell stimulation.[32] Since glycerophospholipids contain two
aliphatic chains at both the *sn*-1 and *sn*-2 positions of the glycerol moiety, it is also important
to know the fatty acid composition of individual molecular species. Table 6 shows the
molecular species composition of diacyl glycerophosphocholine, diacyl glycerophosphoethanolamine, alk-1-enylacyl glycerophosphoethanolamine, diacyl glycerophosphoserine, and
diacyl glycerophosphoinositol of rat erythrocytes.[12] It is interesting to note that the diacyl
glycerophosphoethanolamine fraction contained a much higher proportion of unsaturated
fatty acid species (about 4.3 double bonds per molecule of phospholipid) than the diacyl

TABLE 6
PHOSPHOLIPID MOLECULAR SPECIES OF RAT
ERYTHROCYTES

	Percent of total				
Molecular species	Diacyl-GPC	Diacyl-GPE	Alkenyl-cyl-GPE	Diacyl-GPS	Diacyl-GPI
22:6/20:4	nd	1.37	nd	5.60	nd
20:4/20:4	0.13	4.12	nd	21.99	nd
18:2/20:4	0.43	5.13	0.83	8.02	nd
18:1/20:5	0.37	3.58	2.05	1.16	nd
16:0/20:5	nd	nd	0.23	nd	nd
16:0/22:6	2.46	3.16	5.20	0.75	0.76
18:2/18:2	nd	1.55	nd	nd	nd
18:1/22:5	nd	1.89	1.88	nd	nd
18:1/20:4	0.79	18.96	12.95	6.94	3.81
16:0/22:5	0.47	1.67	8.36	nd	1.01
16:0/20:4	7.82	16.67	18.15	2.85	19.03
18:0/20:5	nd	nd	0.08	nd	nd
18:0/22:6	1.41	1.37	6.51	5.73	2.95
18:1/22:4	nd	0.95	1.40	nd	nd
18:1/18:2	1.20	6.23	nd	1.55	1.53

glycerophosphocholine fraction (about 2.0 double bonds per molecule of phospholipid). Furthermore, 30% of the molecules of diacyl glycerophosphocholine were disaturated. Table 6 also shows the existence of diunsaturated molecules of phospholipids. Most interesting is the existence of highly unsaturated species like 22:6,20:4; 20:4,20:4; and 18:2,20:4 glycerophospholipids. The 22:6,20:4 and the 20:4,20:4 species of diacyl glycerophosphoethanolamine constituted 1.4 and 4.1% of the total species of this lipid, respectively. The alk-1-enylacyl glycerophosphoethanolamine subclass contained predominantly polyunsaturated fatty acids at position 2, with 97% of the acyl moieties containing at least four double bonds. Diacyl glycerophosphoserine and diacyl glycerophosphoinositol also contained large amounts of highly unsaturated fatty acids. These results show the existence of specific molecules of glycerophospholipids which are highly unsaturated. Ether glycerophospholipids are particularly rich in polyunsaturated fatty acids, suggesting a role as storage reservoirs for these fatty acids. This might be due to the apparent protective nature of the ether bond against hydrolysis of the acyl group at the *sn-2* position by phospholipase A_2.

E. BACTERIAL ETHER LIPIDS

Over 11 different basic ether lipid structures exist in bacteria. Ether lipids are major constituents of the membrane of a variety of anaerobic eubacteria and archaebacteria, many of which live in extreme environments.[25] The characteristic ether lipids of anaerobic eubacteria are alk-1-enylacyl glycerolipids having as polar headgroups ethanolamine, glycerol, serine, *N*-monomethylethanolamine, and the plasmalogen form of cardiolipin. In addition to these "classical" forms of plasmalogens, bacteria also contain glycerol acetal forms of the major plasmalogens[26] and dimeric plasmalogens (named diabolic acids) cross-linked by very long-chain dicarboxilic fatty acids[27] (Figure 1). Alkyl ethers are present in small amounts in anaerobics. In addition to the "classical" 1-alkyl-2-acyl form, dialkyl glycerophospholipids have been identified in the anaerobic eubacteria *Thermodesulfotobacterium commune*.[28] Archaebacteria possess membranes composed almost exclusively of dialkyl glycerolipids.[29] Diethers containing two C_{20} phytanyl chains are found in all archaebacteria. Thermophilic and methanogenic archaebacteria contain diglycerol tetraethers containing two C_{40} phytanyl chains (Figure 1). These tetraethers often occur as phosphoglycerolipids in which sugars and phosphate groups are linked to opposite halves of the tetraether molecule.[29,30]

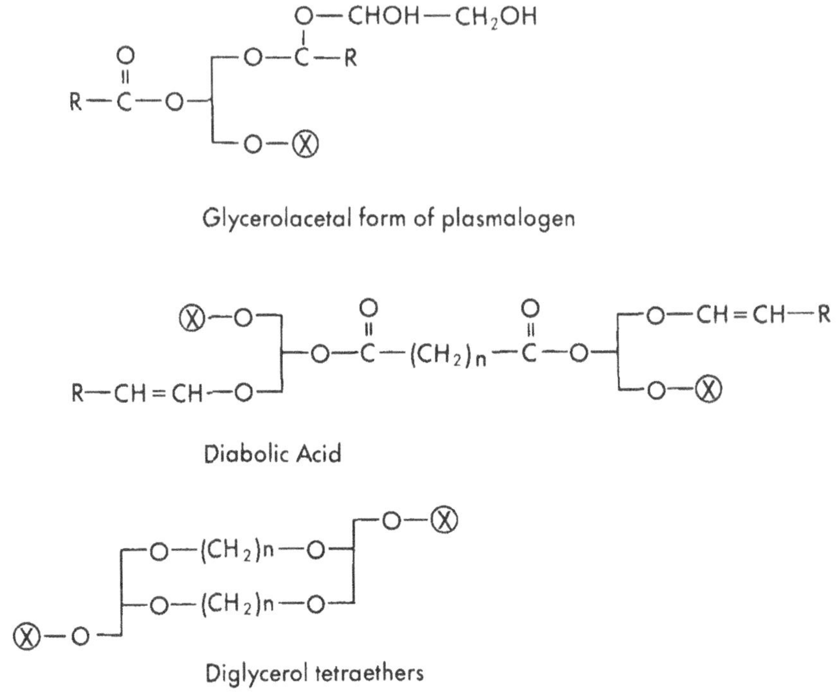

Glycerolacetal form of plasmalogen

Diabolic Acid

Diglycerol tetraethers

FIGURE 1.   Unusual ether lipids found in bacteria. n, Variable carbon number of alkyl chains; X, variable polar headgroup.

These unusual ether lipids in bacteria seem to play specific roles in maintaining membrane stability in these organisms, many of which are extreme halophiles, thermoacidophiles, and methanogens. Since these bacteria live in environments which are similar to the primitive Earth's biosphere, these ether lipid structures might be surviving traces of ancient lipid structures. Recently, bisglyceryl ether lipids, with the general structure 1-alkyl-2-acyl-3-(2',3'-diacylglycerol)-glycerol, have been identified in Harderian gland tumors of mice.[33] The possible biological function of this novel class of lipids is unknown.

REFERENCES

1. **Barenholz, Y. and Thompson, T. E.,** Sphingomyelins in bilayers and biological membranes, *Biochim. Biophys. Acta.* 604, 129, 1980.
2. **White, D. A.,** The phospholipid composition of mammalian tissues, in *Form and Function of Phospholipids,* Ansell, G., Hawthorne, J. N., and Dawson, R. M. C., Eds., Elsevier, Amsterdam, 1973, 441.
3. **Bell, R. M. and Coleman, R. A.,** Enzymes of glycolipid synthesis in eukaryotes, *Annu. Rev. Biochem.,* 49, 459, 1980.
4. **Ansell, G. B. and Spenner, S.,** Phosphatidylserine, phosphatidylethanolamine and phosphatidylcholine, in *Phospholipids,* Hawthorne, J. N. and Ansell, G. B., Eds., Elsevier, Amsterdam, 1982, 1.
5. **Ray, T. K., Shipski, V. P., Barclay, M., Essner, E., and Archibald, F. M.,** Lipid composition of rat liver plasma membranes, *J. Biol. Chem.,* 244, 5528, 1969.
6. **Davis, J. T., Bridges, R. B., and Coniglio, J. C.,** Changes in lipid composition of the maturing rat testis, *Biochem. J.,* 1966, 98, 342.
7. **Simon, G. and Rouser, G.,** Species variations in phospholipid class distribution of organs: heart and skeletal muscle, *Lipids,* 4, 607, 1969.
8. **McMurray, W. C.,** Phospholipids in subcellular organelles and membranes, in *Form and Function of Phospholipids,* Ansell, G. B., Hawthorn, J. N., and Dawson, R. M. C., Eds., Elsevier, New York, 1973, 205.

9. **Keenan, T. W. and Morrey, D. J.**, Phospholipid class and fatty acid compositin of Golgi apparatus isolated from rat liver and comparison with other cell fractions, *Biochemistry*, 9, 19, 1970.
10. **McMurray, W. C. and Magee, W. L.**, Phospholipid metabolism, *Annu. Rev. Biochem.*, 41, 129, 1972.
11. **Quinn, P. J.**, *The Molecular Biology of Cell Membranes*, 1st ed., MacMillan, London, 1976, chap. 1.
12. **Esko, J. D. and Raetz, C. R. H.**, Synthesis of phospholipids in animal cells, in *The Enzymes*, Vol. 16, Boyer, P. B., Ed., Academic Press, New York, 1983, 207.
13. **Robinson, M., Blank, M. L., and Snyder, F.**, Highly unsaturated phospholipid molecular species of rat erythrocyte membranes: selective incorporation of arachidonic acid into phosphoglycerides containing polyunsaturation in both acyl chains, *Arch. Biochem. Biophys.*, 250, 271, 1986.
14. **Blank, M. L., Spector, A. A., Kaduce, T. L., and Snyder, F.**, Composition and incorporation of [^{3}H] arachidonic acid into molecular species of phospholipid classes by cultured endothelial cells, *Biochim. Biophys. Acta*, 877, 211, 1986.
15. **Edgar, M. L., Cress, E. A., and Snyder, F.**, Separation and quantitation of phospholipid subclasses as their diradylglycerobenzoate derivatives by normal-phase high-performance liquid chromatography, *J. Chromatogr.*, 392, 421, 1987.
16. **Getz, G. S., Bartley, W., Stirpe, F., Notton, B. M., and Renshaw, A.**, The lipid composition of rat liver, *Biochem. J.*, 80, 176, 1961.
17. **Menzel, D. B. and Olcott, H. S.**, Positional distribution of fatty acids in fish and other animal lecithins, *Biochem. Biophys. Acta.*, 84, 133, 1964.
18. **Balint, J. A., Beeler, D. A., Treble, D. H., and Spitzer, H. L.**, Studies in the biosynthesis of hepatic and biliary lecithins, *J. Lipid Res.*, 8, 486, 1967.
19. **Kanoh, H.**, Biosynthesis of molecular species of phosphatidylcholine and phosphatidylethanolamine from radioactive precursors in rat liver slices, *Biochem. Biophys. Acta*, 176, 756, 1969.
20. **Possmayer, F., Scherphof, G. L., Dubbelman, T. M. A. R., van Golde, L. M. G., and van Deenen, L. L. M.**, Positional specificity of saturated and unsaturated fatty acids in phosphatidic acid from rat liver, *Biochim. Biophys. Acta*, 176, 95, 1969.
21. **Akesson, B.**, Composition of rat liver triacylglycerols and diacylglycerols, *Eur. J. Biochem.*, 9, 463, 1969.
22. **Fex, G.**, Metabolism of phosphatidycholine, phosphatidylethanolamine and sphingomyelin in regenerating rat liver, *Biochim. Biophys. Acta*, 231, 161, 1971.
23. **Thompson, W. and MacDonald, G.**, Synthesis of molecular classes of cytidine diphosphate diglyceride by rat liver in vivo and in vitro, *Can. J. Biochem.*, 55, 1153, 1977.
24. **Chilton, F. H. and Connell, T. R.**, 1-ether-linked phosphoglycerides are major endogenous sources of arachidonate in the human neutrophil, *J. Biol. Chem.*, 263, 5260, 1988.
25. **Goldfine, H. and Langworthy, T. A.**, A growing interest in bacterial ether lipids, *Trends Biochem. Sci.*, 13, 217, 1988.
26. **Johnston, N. C. and Goldfine, H.**, Phospholipid aliphatic chain composition modulates lipid class composition but not lipid asymmetry in *Clostridium butyricum*, *Biochim. Biophys. Acta.* 813, 10, 1985.
27. **Clark, N. G., Hazlewood, G. P., and Dawson, R. M. C.**, Structure of diabolic acid containing phospholipids isolated from *Butyrivibrio sp. Biochem. J.*, 191, 561, 1980.
28. **Langworthy, T. A., Holzer, G., Zeikus, J. G., and Tornabene, T. G.**, Iso- and antesio-branched glycerol diethers of the thermophilic anaerobe *Thermodesulfotobacterium commune*, *Syst. Appl. Microbiol.*, 4, 1, 1983.
29. **DeRosa, M., Gambacorta, A., and Gliossi, A.**, Structure, biosynthesis, and physicochemical properties of archaebacterial lipids, *Microbiol. Rev.*, 50, 70, 1986.
30. **Hayashi, Y., Urade, R., and Kito, M.**, Distribution of phospholipid molecular species containing arachidonic acid and cholesterol in V79-UF cells, *Biochim. Biophys. Acta*, 918, 267, 1987.
31. **Lange, Y., Swaisgood, M. H., Ramos, B. V., and Steck, T. L.**, Plasma membranes contain half the phospholipid and 90% of the cholesterol and sphingomyelin in cultured human fibroblasts, *J. Biol. Chem.*, 264, 3786, 1989.
32. **Ford, D. A. and Gross, R. W.**, Plasmenylethanolamine is the major store depot for arachidonic acid in rabbit vascular smooth muscle and is rapidly hydrolysed after angiotensin II stimulation, *Proc. Natl. Acad. Sci. U.S.A.*, 86, 3479, 1989.
33. **Kasama, K., Blank, M. L., and Snyder, F.**, Identification of 1-alkyl-2-acyl-3-(2',3' diacylglycerol)glycerols, a new type of lipid class, in Harderian gland tumors of mice, *J. Biol. Chem.*, 264, 9453, 1989.

Chapter 3

PHOSPHOLIPID ORGANIZATION IN CELLULAR MEMBRANES

José M. Mato

TABLE OF CONTENTS

I. PROPERTIES OF MEMBRANE PHOSPHOLIPIDS

In the fluid mosaic model, the prevailing model for membrane structure, phospholipids form a bilayer into which proteins are partially or fully immersed (Figure 1).[1] Phospholipids, as proteins, are not, however, static molecules and are able to perform various types of movements in the bilayer (translational, rotational, and vibrational movements). Phospholipids possess the special property of undergoing phase transitions between "liquid crystalline" and "crystalline" states. Even in the most mobile phospholipid bilayer, a certain order is maintained (i.e., only the polar headgroups of the lipid molecules face the water phase), and this state is termed "liquid crystalline" (fluid). Lowering the temperature decreases the mobility of the phospholipid molecules, and the bilayer achieves a more ordered structure termed "crystalline" (solid) (Figure 2). Phase transition in pure phospholipids occurs over a small temperature range, the midpoint of which is called the "transition temperature" (T_m). Above this T_m, the motion of the lipid chain increases, showing flexing and twisting of the methylene groups and a marked oscillation and rotation of the methyl group at the end of the fatty acid chain.[2] In addition, the polar headgroup of the lipid molecules increase their mobility[3] and the phospholipid molecules diffuse.[4]

In the crystalline state, the acyl chains of the phospholipids are fully extended, with their bonds in the all-*trans* conformation. In the fluid state, gauche isomers around the C-C bonds of the acyl chain occur, producing short-lived "kinks" and decreasing the order of the bilayer.[5] *Cis* double bonds are analogous to permanent kinks and disrupt the packing of the phospholipids in the solid state, favoring the fluid structure of the bilayer.[5]

The T_m of a membrane varies with the phospholipid composition of the bilayer and with the presence of other molecules (e.g., cholesterol, proteins, degree of hydration, etc.).[6-9] As a rule, for a given headgroup, decreasing the fatty acid chain length or the introduction of *cis* double bonds lowers the T_m. Thus, whereas the T_m of 1,2-dipalmitoyl phosphatidylcholine is 41.4°C, that of distearoyl phosphatidylcholine is 54.9°C and that of 1,2-dioleoyl phosphatidylcholine, −22°C.[7,10] The T_m of saturated phosphatidylcholines does not increase linearly as a function of chain length. The pattern of the influence of lipid chain length on T_m is similar for the various glycerophospholipids, so that the effect of chain length on T_m is independent of the headgroup.[11] Phospholipids occurring in nature most often have one saturated and one unsaturated fatty acid per molecule. These phospholipids have sharp phase transitions at a temperature which is halfway between the phase transition for the respective phospholipids with both fatty chains identical. The T_m also varies with the nature of the polar headgroup. For a given fatty acid composition, phosphatidylcholine molecules have a lower T_m than the corresponding phosphatidylethanolamine analogs. Thus, whereas 1,2-dipalmitoyl and 1,2-dimiristoyl phosphatidylcholine have a T_m of 41.4 and 23.9°C, respectively, 1,2-dipalmitoyl and 1,2-dimiristoyl phosphatidylethanolamine have a T_m of 63.1 and 49.5°C, respectively.[10-12] This is due to charge repulsion between adjacent phospholipid headgroups that cause a lateral expansion of the phospholipid bilayer favoring the fluid state. This repulsion varies with the different polar headgroups. For a given chain length, phosphatidylcholine and the sodium salt of phosphatidylglycerol have the lowest T_m. Near neutral pH, all other phospholipids have a higher T_m. Protonation of the headgroup increases the T_m of the acidic phospholipids (phosphatidic acid, phosphatidylglycerol, phosphatidylinositol, and phosphatidylserine). Cardiolipin has the highest T_m of all phospholipids at neutral pH.[11] Methylation of the ammonium group of phosphatidylethanolamine to form the mono and dimethyl derivatives results in a progressive change in the T_m of the corresponding phospholipids. The T_m is increased by an average of 7.7 to 8.7°C per methyl group incorporated.[13,14] As phosphatidylcholine biosynthesis can occur by sequential methylation of the amino group of phosphatidylethanolamine, headgroup modification might provide a mechanism for homeoviscous control of the membrane (see Chapter 7). Ether lipids have, in

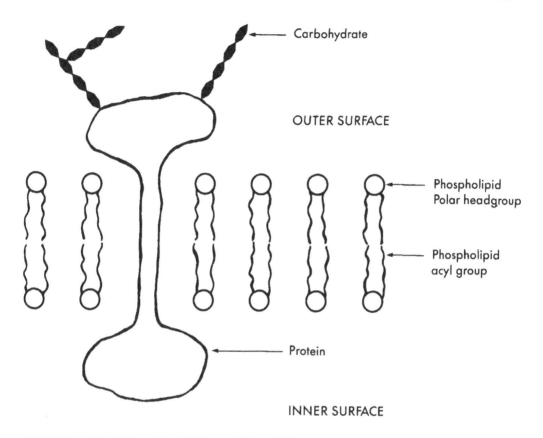

FIGURE 1. In the fluid mosaic model, the prevailing model for membrane structure, phospholipids form a bilayer into which proteins are partially or fully immersed.

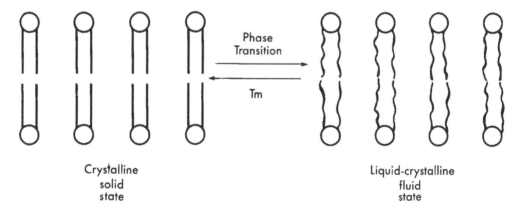

FIGURE 2. Phase transition of phospholipids. The phospholipids exist in two forms: (1) a solid or crystalline state, in which the acyl chains are fully extended in the all-*trans* conformation and the lipid molecules pack closely together, and (2) a fluid or liquid-crystalline state, in which the acyl molecules undergo flexing and twisting of the methylene groups, leading to formation of *gauche* isomers. In the fluid state, the phospholipid molecules gain fast lateral and rotational diffusion.

general, T_ms that are some 2 to 6°C higher than the corresponding diacyl analogs and are therefore more rigid.[11,13,15] These differences are thought to arise from alterations in both chain packing and the intermolecular hydrogen-bonding arrangement as a consequence of the ether vs. ester chain linkages. The T_m of sphingomyelin-containing liposomes is similar to that of the corresponding phosphatidylcholine. Thus, N-palmitoylsphingosine phosphor-

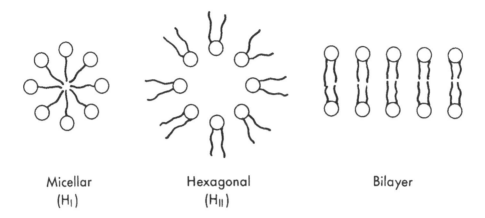

Micellar
(H$_I$)

Hexagonal
(H$_{II}$)

Bilayer

FIGURE 3.   Structures of membrane phospholipid aggregates.

ylcholine and 1,2-dipalmitoyl phosphatidylcholine have a T$_m$ of 41.3 and 41.1°C, respectively.[16] As mentioned above, the insertion of a *cis* double bond lowers the T$_m$ of phospholipids. There is, however, no direct correlation between the number of double bonds and the magnitude of the decrease in the T$_m$. Thus, the T$_m$ of 1-stearoyl, 2-linolenoyl, 1-stearoyl,2-linolenoyl, and 1-stearoyl,2-arachidonoyl phosphatidylcholine are -18, -13, and -12.6°C, respectively.[17] A correlation might exist, however, for phospholipids containing two unsaturated fatty acids, but not in the case of the most commonly occurring lipids with one saturated and one unsaturated species. Positional isomers of saturated 1,2-diacyl phosphatidylcholine have different T$_m$s. Thus, the T$_m$ of 1-myristoyl,2-palmitoyl phosphatidylcholine and 1-palmitoyl,2-myristoyl phosphatidylcholine are 35.5 and 27.2°C, respectively.[11] In these positional isomers, the orientations of the two fatty acid chains near the glycerol backbone are different, so that a given chain at position 1 has a greater bilayer penetration than at position 2.[18]

The area occupied by the various phospholipids that form a membrane is also a function of the polar headgroup, fatty acid length, double bonds, and the presence of other molecules (e.g., cholesterol, ions, and pH). Thus, whereas the surface area of 1,2-dimyristoyl phosphatidylcholine is 72 Å^2 per molecule, the area occupied by 1,2-dipalmitoyl phosphatidylcholine and 1,2-dimyristoyl phosphatidylethanolamine is 58 and 50 Å^2 per molecule, respectively.[19,20] Cholesterol increases the packing area of phospholipids by adding a hydrophobic "spacer" between two lipid molecules and has a "filling" role, as shown by the observation that mixtures of lysophosphatidylcholine plus cholesterol form bilayers, whereas neither lipid alone does so.[21]

II. PHASE PROPERTIES OF MEMBRANE PHOSPHOLIPIDS

Individual species of biological membrane phospholipids adopt in their liquid crystalline state one of three structures: micellar (H$_I$), hexagonal (H$_{II}$), or bilayer (Figure 3). A simple shape concept model has been proposed to predict the type of structure that phospholipids will adopt upon hydration.[82] According to this model, lipid bilayers are formed by hydrated phospholipids that are approximately cylindrical in shape as a result of a balance between the sizes of the hydrated polar headgroup and the volume occupied by the acyl chains. Phospholipids with a small headgroup, compared to the volume of the acyl chains, will have a conical shape and will tend to form hexagonal structures. An increase in the effective polar headgroup area, as occurs in the conversion of phosphatidylethanolamine to phosphatidylcholine or monoglycosyldiacylglycerol to diglycosyldiacylglycerol, serves to stabilize the bilayer structure. Similarly, conversion of plasmenylethanolamine to its glycerol acetal also serves to stabilize the bilayer structure, a process which might be important for the regulation of bilayer stability in *Clostridium butyricum* membranes.[82] At physiological conditions, the micellar structure is preferred by lysophospholipids, which have a large headgroup in comparison to the alkyl chain size.[22] At physiological conditions, all phosphatidylcholine molecules adopt a lamellar structure, and at higher temperatures, or low hydration, unsaturated phosphatidylcholine molecules can also adopt nonlamellar hexagonal H_{11} structures.[23-25] Phosphatidylethanolamine molecules at physiological temperatures adopt hexagonal H_{11} structures. Lamellar and hexagonal phosphatidylethanolamine structures can be observed, depending on the temperature, hydration, ionic strength, pH, acyl chain length, the presence of other lipid molecules, etc.[25-27,91] At physiological pH, phosphatidylserine, phosphatidylinositol, and phosphatidic acid all carry a net negative charge and they all adopt a lamellar structure.[28-30] Sphingomyelin adopts a lamellar structure over a large range of temperatures.[31]

In summary, under appropriate conditions (e.g., pH, temperature, and ionic strength), practically all phospholipids found in biological membranes can adopt either a liquid crystalline bilayer or a hexagonal H_{11} structure. Physiological signals might therefore be able to trigger changes from lamellar to hexagonal H_{11} structures (or vice versa) in a limited number of phospholipids within a biomembrane, and this might play a role in processes such as membrane fusion. However, it is important to note that for some investigators, the formation of nonlamellar lipid structures in biological membranes is unlikely, and that the role of nonbilayer-forming lipids might be to impart some special properties to the lipid bilayer phase.[90]

III. PHASE PROPERTIES OF PHOSPHOLIPID MIXTURES

Biomembranes are complex mixtures of phospholipid molecules having different headgroups and/or acyl chains which are surrounded by other molecules (i.e., proteins, cholesterol, and carbohydrates). As a result of all this complexity, biomembranes do not undergo well-defined lipid phase transitions over a narrow temperature range in which all the phospholipid molecules move from a crystalline to a fluid state. On the contrary, by lowering the temperature, biomembranes undergo "lipid phase separations". This term indicates a situation in which several types of lipid phases with different lipid compositions and structures coexist within the same biomembrane. An approach to study this phenomenon is to determine the phase structure behavior of binary mixtures of phospholipids. When both phospholipids have one type of acyl chain ("homoacid"), the same polar headgroup, and similar acyl chains (i.e., 1,2-dimyristoyl and 1,2-dipalmitoyl phosphatidylcholine), the lipid mixture is close to ideal and a relatively simple phase diagram is obtained.[32] There are only three phases (fluid, fluid plus solid, and solid) and no separation between both phospholipid species exists in either phase, indicating that there is a random organization of the headgroups. As mentioned in Chapter 2, biomembrane phospholipids most often contain two different types of fatty acid chains ("heteroacid"). Mixtures of a "homoacid" phospholipid with a heteroacid phospholipid, each having the same headgroup, show a very different solid-state pattern. These mixtures are characterized by the yield of a nonhomogeneous solid phase containing areas with each of the phospholipid components, as well as by the existence of solid phospholipid mixtures.[33] The greater the difference in the T_m, the less miscible are the components in the gel. In the liquid crystalline phase, these types of mixtures are more nearly ideally mixed than in the solid phase. Immiscibility in the fluid state is more rare; however, certain phospholipid mixtures have been shown to be partially immiscible in the fluid phase.[33,34] Solid-state immiscibility is also observed when two phospholipids with different polar headgroups are mixed. As a rule, "lamellar" lipids, such as phosphatidylcholine, are able to stabilize hexagonal H_{11} lipids, such as phosphatidylethanolamine, indicating ideal mixing properties in the fluid state. For binary systems of phosphatidylcholine, miscibility in the gel state is dependent on the chain length difference of the two phosphatidylcholine species. If the chain length difference is less than four carbons, solid-phase miscibility is found at all lipid ratios. However, if the difference is greater than four carbons, immiscibility occurs by the separation of the phosphatidylcholine molecules with longer acyl chains.[76,83] For binary mixtures of ether- and ester-linked phospholipids, miscibility is found in the fluid state, however, below T_m, lipid immiscibility with areas containing each of the phospholipid species is observed.[77] The phase behavior of cerebroside with phosphatidylcholine has been determined by calorimetric studies.[35] Whereas, at low mole fractions, cerebroside/phosphatidylcholine mixtures were found to be miscible, at high mole fractions, the mixture behaved nonideally. The nonideal behavior seems to be due to different interactions of the two cerebroside fractions, kerasin (β-D-galactosyl-N-acyl-D-sphingosine) and phrenosin [(β-D-galactosyl-N-(2-D-hydroxyacyl)-D-sphingosine)], with phosphatidylcholine. These different abilities to mix with phosphatidylcholine of kerasin and phrenosin might be of physiological importance in myelin, where cerebroside is a major lipid component, and in various plasma membranes. Certain neutral glycosphingolipids form nonrandom dispersions (gel-like microdomains) in liquid crystalline phosphatidylcholine bilayer systems.[36-39] In contrast to this situation, negatively charged glycosphingolipids, such as ganglioside GM_1, are molecularly dispersed in phosphatidylcholine bilayers.[40] The modulatory role of N-acetylneuraminic acid, which confers the negative charge to GM_1, may have physiological importance. First, although glycosphingolipids are minor components of mammalian cell membranes since they are confined to the external surface of the cell, in this surface they might be major components (see Chapter 4). Second, specific glycosphingolipids have been shown to serve as receptors for toxins, viruses, and some hormones, as well as to act as antigenic determinants and mediate immune responses and cellular proliferation (also see Chapter 9).[86] Third, specific sialidases are thought to be localized in the plasma membrane. It is therefore conceivable that signals that stimulate removal of sialic acid from plasma membrane glycolipids might promote lateral rearrangements, with formation of microdomains within the cell surface that may function as specific recognition sites, analogous to the carbohydrate domains of membrane glycoproteins. In fact, sialylation and desialylation are known to be important factors in the clearance of erythrocytes and lymphocytes from the circulation.[41]

IV. EFFECT OF CHOLESTEROL ON PHOSPHOLIPID PHASE TRANSITION

Cholesterol, which is found almost exclusively in the plasma membrane of mammalian cells, is a most important component of these membranes, where it accounts for about 30 to 50% of the total lipid. Cholesterol consists of a planar hydrophobic sterol ring which has, at one end, a hydroxyl group which constitutes the polar moiety of the molecule, and, at the other end, a hydrocarbon chain. Cholesterol is inserted into the lipid bilayer with its polar hydroxyl group facing the polar region of the membrane and the hydrocarbon moiety inserted in the apolar area (Figure 4). Whether the β-hydroxyl group of cholesterol may participate in hydrogen bonding, with the oxygen forming the carbonyl group that links the fatty acid chain to the glycerol backbone (or some group in the headgroup), has not been demonstrated unequivocally.[42-44] The sterol ring interacts with the adjacent fatty acid chains, reducing their mobility and producing a condensation effect on the acyl chains of the bilayer in the fluid state. Conversely, the insertion of the hydroxyl group between two adjacent phospholipid molecules increases the separation between the headgroups, causing an increase in the mobility of the polar terminal moiety of the phospholipid. As a result, cholesterol decreases the mobility of the acyl chain core of fluid lipid bilayers, yielding a more rigid,

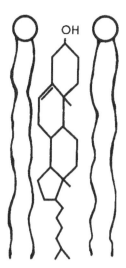

FIGURE 4.   Insertion of cholesterol into a phospholipid bilayer.

ordered structure. Infrared spectroscopic results have shown that the addition of cholesterol to various membranes lipids causes a reduction of the number of gauche conformers in the fluid phase of the lipids, and that this solidifying effect increases with cholesterol concentration.

In direct contrast with the ordering effect of cholesterol in fluid state lipid bilayers, cholesterol increases fluidity in the solid state. In the solid state, phospholipids do not pack together, excluding the cholesterol molecules, but, rather, pack with cholesterol molecules inserted between them, leading to an increase in bilayer fluidity. This dual effect of cholesterol on fluidity, depending on whether the bilayer is in the solid or fluid state, leads to smearing of the lipid phase transition, inhibiting the formation of the crystalline state and decreasing the permeability of the membrane to water and small molecules.[45,46] In the solid state, positional isomers of unsaturated phosphatidylcholines interact differently with cholesterol.[47,48] Thus, cholesterol causes a greater reduction in the T_m of 1-oleoyl-2-stearoyl phosphatidylcholine than in the isomer 1-stearoyl-2-oleoyl phosphatidylcholine.[48] This point is of interest in view of the fact that hepatomas have a high content of 1-unsaturated-2-saturated and diunsaturated phosphatidylcholines and cholesterol.[49] Cholesterol also has the ability to reduce the steric hindrace of the phosphatidylcholine headgroup, and this might have implications in terms of the close approach of both adjacent bilayers and small hydrophobic molecules during, for example, vesicle adhesion and membrane fusion.[88] Similarly, diacylglycerol, at concentrations which are likely to be generated during activation of the phosphatidylinositol cycle (see Chapter 7), can lower the lamellar-hexagonal phase transition temperature and the temperature for fast membrane fusion by 15 to 20°C, suggesting that diacylglycerol increases the susceptibility of phospholipid membrane fusion.[92]

V. EFFECTS OF DOLICHOLS AND NEUTRAL LIPIDS ON PHOSPHOLIPID ORGANIZATION IN MEMBRANES

Dolichols comprise a family of long-chain polyisoprenoid alcohols (typically 18 to 20 isoprene units) (Figure 5) which are present in most eukaryotic membranes and play a role in glycosylation of proteins by serving as lipid carriers of preassembled oligosaccharide chains.[50,81,87] The hydroxyl group of dolichol is often esterified with fatty acids, and a small proportion is phosphorylated. The chain length of this unusual lipid exceeds twice the thickness of the phospholipid bilayer. This fact, together with its marked hydrophobicity,

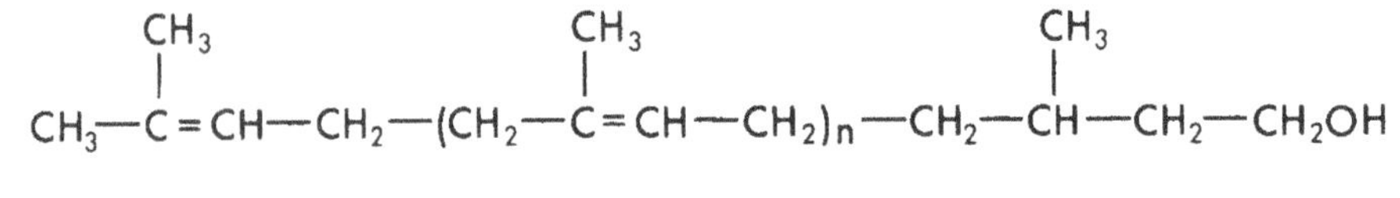

$$CH_3-\underset{\underset{CH_3}{|}}{C}=CH-CH_2-(CH_2-\underset{\underset{CH_3}{|}}{C}=CH-CH_2)_n-CH_2-\underset{\underset{CH_3}{|}}{CH}-CH_2-CH_2OH$$

FIGURE 5.   Structure of dolichol.

suggests that dolichols modulate the organization of the phospholipids in the membrane. Dolichol, as well as dolichol esters, do not affect the phase transition in phosphatidylcholines, indicating that these polyprenols do not mix with phosphatidylcholine bilayers.[50] In contrast, the introduction of negative charges in the dolichol molecule by addition of a phosphate group abolishes the solid to liquid-crystalline transition in phosphatidylcholine bilayers.[51,52] In the case of phosphatidylethanolamine, dolichol, and dolichol phosphate, but not dolichol esters, mix well with phosphatidylethanolamine in the liquid-crystalline state, increasing its fluidity.[51,53,89] Both polyprenols have a destabilizing effect on phosphatidylethanolamine containing bilayers, resulting in the formation of H_{11} structures and lipidic particles, enhancing the fusion between large unilamellar vesicles.[51,53] The effects of dolichol on the motion of phospholipid molecules has also been investigated, using spin-labeled probes. The results of these experiments show that dolichol reduces the motion of the spin probes at carbon 16, suggesting that polyprenols penetrate into the lipid core region of the bilayer.[54] The observation that dolichol phosphate induces membrane disorder led to the proposal that such an effect might facilitate transmembrane movement of the dolichol-linked oligosaccharides in natural membranes. Interestingly, mannosylphosphodolichol synthase activity in microsomal membranes from rat parotid gland is activated under conditions that increase cAMP-mediated protein phosphorylation of the membranes, supporting the view that extracellular signals regulate dolichol phosphate metabolism.[93]

Plasma membranes from a human leukemic T lymphoblast, rat mammary adenocarcinoma cells, and Chinese hamster ovary cells have been found to contain triglycerides (up to 6% of the total lipid content).[55,56] Plasma membrane triglycerides tumble isotropically in domains that are not in diffusible exchange with neighboring phospholipids, forming a sphere that resembles lipoprotein particles.[55-57] Based on these results, a model has been proposed for malignant plasma membranes in which spheres of monolayer phospholipid enclosing a core of tumbling neutral lipids are intercalated in the bilayer.[56]

VI. MOVEMENTS OF MEMBRANE PHOSPHOLIPIDS

In the fluid state, individual phospholipid molecules undergo rapid rotational diffusion around the axis of the phospholipid and rocking motions (Figure 6). In addition, phospholipids can migrate laterally in the plane of the lipid crystalline bilayer (Figure 6). This lateral diffusion of phospholipid molecules is also fast, with diffusion coefficients about 10^{-8} cm^2/s.[7] A third type of motion of membrane phospholipids are the intrachain movements of the fatty acid chains. In the fluid state, the fatty acid chains are constantly rotating around the C-C bonds, forming "kinks", and flexing the acyl chains (Figure 6). These factors are the main contributors to the viscosity of the membrane, which is about 100 times more viscous than water.[7,57] The reciprocal of the membrane viscosity, termed "membrane fluidity", is proportional to the rotational and lateral diffusion rates of the membrane components.[59] Fluidity is very low when the phospholipids are in the solid state and increases enormously at the phase transition temperature. Above the T_m, there is a marked increase in the intra- and intermolecular motion of the phospholipids. The individual phospholipid molecules have increased intrachain movements and there is a decrease in the chain-order parameters. Above the T_m, there is an increase in intermolecular motion, which yields a decreased resistance to diffusion and, consequently, increased fluidity. There is a relationship between the vis-

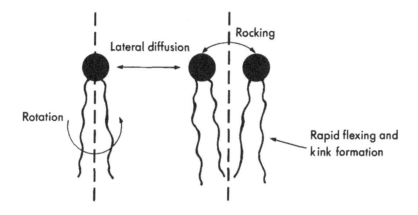

FIGURE 6.   Movements of membrane phospholipids.

cosity of the membrane at a fixed temperature and the T_m; however, this relationship is not a simple one, and also depends on phospholipid composition and lipid-lipid and lipid-protein interactions.[7,11,58,59]

For a given phospholipid molecule, the degree of motion of the various regions of the molecule varies, depending on its position in the fluid lipid bilayer. The mobility of particular carbon atoms in a molecule phospholipid can be studied by incorporating [^{13}C] at specific positions and studying its behavior. The glycerol backbone of a phospholipid, being the mobile parts of the polar headgroup and the acyl chains, is more rigid than the rest of the molecule. There is an increase in motion along the acyl chain, which is maximal at the terminal methyl group.[60] The rotational diffusion of the polar headgroup of a given phospholipid molecule is lower than that observed for the rest of the molecule.[79] There is, therefore, a mobility as well as a fluidity gradient within the membrane, which is maximal in the core of the bilayer.

It is important to note that in the membrane model of Figure 1, the methyl end groups of the bilayer acyl chains are in contact with each other with implies that the chain length of the acyl groups are the same. This is, however, not the case, since biological membranes contain a large variety of molecular species that differ in their fatty acid composition. Moreover, the glycerol backbone of a given phospholipid is not parallel to the membrane surface, which means that even if the chain length of both fatty acids are the same, the acyl group at position 1 penetrates deeper into the bilayer than that at position 2. To compensate for this difference in fatty acid chain length in membrane phospholipids, it has been proposed that their ends can interdigitate.[85]

Phospholipids can also move from one half of the bilayer to the other half. This transbilayer movement is generally referred to as "flip-flop" (Figure 7). This is a one-by-one exchange process and, thus, the total number of phospholipid molecules in each half remains constant. Transbilayer movements involve the translation across the hydrophobic, apolar core of the bilayer of the polar headgroup. This movement is thermodynamically unfavorable and, therefore, very slow. To determine flip-flop rates, a technique was developed in which a phosphatidylcholine transfer protein is used to introduce labeled phosphatidylcholine molecules into the outer layer of intact erythrocytes.[61] Phosphatidylcholine transfer protein mediates the transport of phosphatidylcholine between donor and acceptor membranes.[84] After removal of the protein and the donor-labeled phosphatidylcholine system, the erythrocytes are incubated at 37°C to allow the radioactive phospholipid to equilibrate between the two halves of the bilayer. At various times, the phosphatidylcholine present at the outer surface of the erythrocyte is converted into its lyso derivative by incubation with phospholipase A_2. The lyso-phosphatidylcholine generated represents the amount of phosphatidyl-

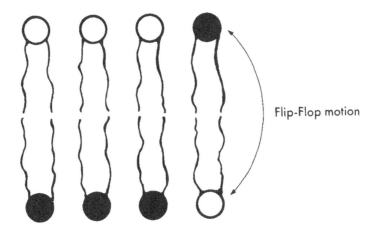

FIGURE 7.   Transbilayer (flip-flop) movement of membrane phospholipids.

choline in the outer surface, and the remaining phosphatidylcholine represents the amount of phospholipid at the internal monolayer. By separation of both labeled lipids, the rate of transbilayer movement can be easily determined. Using this approach, the half-life for transbilayer equilibration in human erythrocytes has been calculated to be 8.1, 10.8, 12.8, and 26.9 h for soybean, rat liver microsomal, egg and dipalmitoyl phosphatidylcholine, respectively.[61,62] These results also indicate that there is a positive correlation between the rate of transbilayer motion and the degree of unsaturation. A variety of studies using phospholipid transfer proteins, electron spin resonance methods, or chemical methods indicate that phospholipid flip-flop in both model and biological membranes is generally a very slow process, with half-lives of hours to days.[62,63,78] The exception to this rule are microsomes, where rapid phospholipid transbilayer movements have been measured.[64-66] This might be due to the presence of proteins in the microsomal membrane which are responsible for inducing nonbilayer structures and rapid phospholipid flip-flop.[62] The occurrence of nonbilayer lipid structures in artificial membranes has been correlated with enhanced transbilayer movements of the phospholipids.[67-69] Enhanced transbilayer mobility of the four major phospholipids (phosphatidylcholine, phosphatidylethanolamine, phosphatidylserine, and sphingomyelin) has been observed using electron spin resonance in *Plasmodium knowlesi* (malaria)-infected monkey erythrocytes.[70] Pathological erythrocytes often show enhanced flip-flop rates. Thus, sickling induces a two- to fourfold increase in phosphatidylcholine flip-flop.[70,71] Similarly, spectrin- and protein 4.1-deficient erythrocytes,[71] and McLeod and Rh null erythrocytes,[72,73] also have enhanced phosphatidylcholine transbilayer rates. The mechanism of this increase in phospholipid flip-flop in pathological erythrocytes is not known, and its physiological implications have not been determined.

The exchange of cholesterol between cholesterol/phospholipid vesicles and an excess of erythrocyte ghosts has also been investigated. Independent of the nature of the phosphatidylcholine used in the vesicles, only the cholesterol associated with the outer surface of the bilayer was exchanged, indicating that, at equilibrium, transbilayer movement of cholesterol is very slow.[69,74,75] Under nonequilibrium conditions, that is, when cholesterol is both exchanged between and depleted from the donor vesicle and the cell ghosts, transbilayer movements can occur.[63,75]

REFERENCES

1. **Singer, S. J. and Nicholson, G. L.,** The fluid mosaic model of the structure of cell membranes, *Science,* 175, 720, 1972.
2. **Chapageman, D.,** Protein-lipid interactions in model and natural biomembranes, in *Biological Membranes,* Chapman, D., Ed., Academic Press, New York, 1982, chap. 4.
3. **Veksli, Z., Salsbury, N. J., and Chapman, D.,** Physical studies of phospholipids. XII. Nuclear magnetic resonance studies of molecular motion in some pure lecithin-water systems, *Biochim. Biophys. Acta,* 183, 434, 1969.
4. **Tilcock, C. P. S.,** Lipid polymorphism, in *Chemistry and Physics of Lipids,* Cullis, P. R. and Hope, M. J., Eds., Elsevier, Ireland, 1986, 40, 109.
5. **Lagaly, G., Weiss, A., and Stuke, E.,** Effect of double-bonds on biomolecular films in membrane models, *Biochim. Biophys. Acta,* 470, 331, 1977.
6. **Chapman, D.,** Phase transitions and fluidity characteristics of lipids and cell membranes, *Q. Rev. Biophys.,* 8, 185, 1975.
7. **Quinn, P. J. and Chapman, D.,** The dynamics of membrane structure, *CRC Crit. Rev. Biochem.,* 8, 1, 1980.
8. **Edidin, M.,** Molecular motions and membrane organization and function, in *Membrane Structure,* Finean, J. B. and Michell, R. H., Eds., Elsevier/North-Holland, Amsterdam, 1981, chap. 2.
9. **Thompson, T. E. and Huang, C.,** Dynamics of lipids in biomembranes, in *Membrane Physiology,* Andreoli, T. E., Hoffman, J. F., and Fanestil, D. D., Eds., Plenum Press, New York, 1978, chap. 2.
10. **Mabrey, S. and Sturtevant, J. M.,** Investigation of phase transitions of lipids and lipid mixtures by high sensitivity differential scanning calorimetry, *Proc. Natl. Acad. Sci. U.S.A.,* 73, 3862, 1976.
11. **Keough, K. M. W. and Davis, P. J.,** Thermal analysis of membranes, in *Membrane Fluidity,* Kates, M. and Manson, L. A., Eds., Plenum Press, New York, 1984, chap. 2.
12. **Wilkinson, D. A. and Nagle, J. R.,** Thermodynamics of lipid bilayers, in *Liposomes: From Physical Structure to Therapeutic Applications,* Knight, C. G., Ed., Elsevier/North-Holland, Amsterdam, 1981, chap. 9.
13. **Vaughan, D. J. and Keough, K. M.,** Changes in phase transition of phosphatidylethanolamine- and phosphatidylcholine-water dispersions induced by small modifications in the headgroup and backbone regions, *FEBS Lett.,* 47, 158, 1974.
14. **Casal, H. L. and Mantsch, H. H.,** The thermotropic phase behaviour of N-methylated dipalmitoyl phosphatidylethanolamines, *Biochim. Biophys. Acta,* 735, 387, 1983.
15. **Paltauf, F.,** Ether lipids in biological and model membranes, in *Ether Lipids: Biochemical and Biomedical Aspects,* Mangold, H. K. and Paltauf, F., Eds., Academic Press, New York, 1983, chap. 17.
16. **Barenholtz, Y. and Thompson, T. E.,** Sphingomyelins in bilayers and biological membranes, *Biochim. Biophys. Acta,* 604, 129, 1980.
17. **Coolbear, K. P., Berde, C. B., and Keough, K. M. W.,** Gel to liquid-crystalline phase transitions of aqueous dispersions of polyunsaturated mixed-acid phosphatidylcholines, *Biochemistry,* 22, 1466, 1983.
18. **Seelig, J. and Seelig, A.,** Lipid conformation in model membranes and biological membranes, *Q. Rev. Biophys.,* 13, 19, 1980.
19. **O'Brien, J. S.,** Cell membranes: composition, structure, function, *J. Theor. Biol.,* 15, 307, 1967.
20. **Philips, M. C. and Chapman, D.,** Monolayer characteristics of saturated 1,2-diacylphosphatidylcholines and phosphatidylethanolamines at the air-water interface, *Biochim. Biophys. Acta,* 163, 301, 1968.
21. **Merrill, A. H. and Nichols, J. W.,** Techniques for studying phospholipid membranes, in *Phospholipids and Cellular Regulation,* Kuo, J. F., Ed., CRC Press, Boca Raton, FL, 1985, chap. 2.
22. **Cullis, P. R., Hope, M. J., de Kruijff, B., Verkleij, A. J., and Tilcock, C. P. S.,** Structural properties and functional roles of phospholipids in biological membranes, in *Phospholipids and Cellular Regulation,* Kuo, J. F., Ed., CRC Press, Boca Raton, FL, 1985, chap. 1.
23. **Small, D.,** Phase equilibria and structure of dry and hydrated egg lecithin, *J. Lipid Res.,* 8, 551, 1967.
24. **Luzatti, V., Gulik-Krzywicki, T., and Tardieu, A.,** Polymorphism of lecithins, *Nature,* 218, 1031, 1968.
25. **Cullis, P. R. and de Kruijff, B.,** Lipid polymorphism and the functional roles of lipids in biological membranes, *Biochim. Biophys. Acta,* 559, 399, 1979.
26. **Hui, S. W., Stewart, T. P., Yeagle, P. L., and Albert, D.,** Bilayer to non-bilayer transitions in mixtures of phosphatidylethanolamine and phosphatidylcholine. Implications for membrane properties, *Arch. Biochem. Biophys.,* 207, 227, 1981.
27. **Marsh, D. and Seddon, J. M.,** Gel-to-inverted hexagonal (L_β-H^{11}) phase transitions in phosphatidylethanolamines and fatty acid phosphatidylcholine mixtures, demonstrated by ^{31}P NMR spectroscopy and X-ray diffraction, *Biochim. Biophys. Acta,* 690, 117, 1982.
28. **Cevc, G., Watts, A., and Marsh, D.,** Titration of the phase transition of phosphatidylserine bilayer membranes, *Biochemistry,* 20, 4955, 1981.

29. **Blume, A. and Eibl, H.,** The influence of charge on bilayer membranes. Calorimetric investigations of phosphatidic acid bilayers, *Biochim. Biophys. Acta*, 558, 13, 1979.
30. **Nayar, R., Schimd, S. L., Hope, M. J., and Cullis, P. R.,** Structural preferences of phosphatidylinositol and phosphatidylinositol-phosphatidylethanolamine model membranes. Influence of Ca^{2+} and pH, *Biochim. Biophys. Acta*, 688, 169, 1982.
31. **Shipley, G. G., Avecilla, L. S., and Small, D. M.,** Phase behaviour and structure of aqueous dispersions of sphingomyelin, *J. Lipid Res.*, 15, 124, 1974.
32. **Shimshick, E. S. and McConnell, H. M.,** Lateral phase separation in phospholipid membranes, *Biochemistry*, 12, 2351, 1973.
33. **Davis, P. J. and Keough, K. M. W.,** Phase diagrams of bilayers of dimyristoyl lecithin plus heteroacid lecithins, *Biochim. Biophys. Acta*, 35, 299, 1984.
34. **Wu, S. H. W. and Connell, H. M.,** Phase separations in phospholipid membranes, *Biochemistry*, 14, 847, 1975.
35. **Bunow, M. R. and Lewin, I. W.,** Phase behaviour of cerebroside and its fractions with phosphatidylcholines: calorimetric studies, *Biochim. Biophys. Acta*, 939, 577, 1988.
36. **Correa-Freire, M. C., Barenholz, Y., and Thompson, T. E.,** Glucocerebroside transfer between phosphatidylcholine bilayers, *Biochemistry*, 21, 1244, 1982.
37. **Correa-Freire, M., Freire, E., Barenholz, Y., Biltonen, R. L., and Thompson, T. E.,** Thermotropic behaviour of monoglucocerebroside-dipalmitoylphosphatidylcholine multilamellar liposomes, *Biochemistry*, 18, 442, 1979.
38. **Brown, R. E., Sugar, I. P., and Thompson, T. E.,** Spontaneous transfer of gangliocerebrosylceramide between phospholipid vesicles, *Biochemistry*, 24, 4082, 1985.
39. **Thompson, T. E., Allietta, M., Brown, R. E., Johnson, M. L., and Tillack, T. W.,** Organization of ganglioside GM_1 in phosphatidylcholine bilayers, *Biochim. Biophys. Acta*, 817, 229, 1985.
40. **Brown, R. E. and Thompson, T. E.,** Spontaneous transfer of ganglioside GM_1 between phospholipid vesicles, *Biochemistry*, 26, 5454, 1987.
41. **Schauer, R.,** Sialic acids and their role as biological masks, *Trends Biochem. Sci.*, 10, 357, 1985.
42. **De Kruyff, B., Demel, R. A., Slotboom, A. J., van Deenen, L. L. M., and Rosenthal, A. F.,** The effect of the polar headgroup on the lipid-cholesterol interaction: a monolayer and differential scanning calorimetry study, *Biochim. Biophys. Acta*, 307, 1, 1973.
43. **Huang, C. H.,** A structural model for the cholesterol-phosphatidylcholine complexes in bilayer membranes, *Lipids*, 12, 348, 1977.
44. **Bittman, R., Clejan, S., Lund-Katz, S., and Philips, M. C.,** Influence of cholesterol on bilayers of esters and ether-linked phospholipids. Permeability and ^{13}C-nuclear magnetic resonance measurement, *Biochim. Biophys. Acta*, 772, 117, 1984.
45. **Ladbroke, B. D., Williams, R. M., and Chapman, D.,** Studies on lecithin-cholesterol-water interactions by differential scanning calorimetry and X-ray diffraction, *Biochim. Biophys. Acta*, 150, 333, 1968.
46. **Demel, R. A. and de Kruijff, B.,** The function of sterols in membranes, *Biochim. Biophys. Acta*, 457, 109, 1976.
47. **Davis, P. J. and Keough, K. M. W.,** Scanning calorimetric studies of aqueous dispersions of bilayers made with cholesterol and a pair of positional isomers of 3-*sn*-phosphatidylcholine, *Biochim. Biophys. Acta*, 778, 305, 1984.
48. **Davis, P. J., Kariel, N., and Keough, K. M. W.,** Gel to liquid-crystalline transitions of aqueous dispersions of positional isomers of a heteroacid unsaturated phosphatidylcholine mixed with epicholesterol and cholesterol, *Biochim. Biophys. Acta*, 856, 395, 1986.
49. **Dyatlovitskaya, E. V., Yanchevskaya, G. V., and Bergelson, L. D.,** Molecular species and membrane forming properties of lecithins in normal liver and hepatoma, *Chem. Phys. Lipids*, 12, 132, 1974.
50. **Tavares, A., Coolbear, T., and Hemming, F. W.,** Increased hepatic dolichol and dolichol phosphate-mediated glycosylation in rats fed cholesterol, *Arch. Biochem. Biophys.*, 207, 427, 1981.
51. **van Duijn, G., Verkleij, A. J., de Kruiff, B., Valtersson, C., Dallner, G., and Chojnacki, T.,** Influence of dolichols on lipid polymorphism in model membranes and the consequences for phospholipid flip-flop and vesicle fusion, *Chem. Scr.*, 27, 95, 1987.
52. **McCloskey, M. A. and Troy, F. A.,** Paramagnetic isoprenoid carrier lipids. II. Dispersion and dynamics in lipid membranes, *Biochemistry*, 19, 2061, 1980.
53. **Valtersson, C., van Duijn, G., Verkleij, A. J., Chojnaki, T., de Kruijff, B., and Dallner, G.,** The influence of dolichol, dolichol esters, and dolichol phosphate on phospholipid polymorphism and fluidity in model membranes, *J. Biol. Chem.*, 260, 2742, 1985.
54. **Lai, C. S. and Schutzbach, J. S.,** Localization of dolichols in phospholipid membranes, *FEBS Lett.*, 203, 153, 1986.
55. **May, G. L., Wright, L., Holmes, K. T., Williams, P. G., Smith, I. C. P., Wright, P. E., Fox, R. M., and Mountford, C. E.,** Assignment of methylene proton resonances in NMR spectra of embryonic and transformed cells to plasma membrane triglyceride, *J. Biol. Chem.*, 261, 3048, 1986.

56. **Mountford, C. E. and Wright, L. C.**, Organization of lipids in the plasma membranes of malignant and stimulated cells: a new model, *Trends Biochem. Sci.*, 13, 172, 1988.
57. **Mountford, C. E., Grossman, G., Reid, G., and Fox, R. M.**, Characterization of transformed cells and tumors by proton nuclear magnetic resonance spectroscopy, *Cancer Res.*, 42, 2270, 1982.
58. **Cherry, R. J.**, Rotational and lateral diffusion of membrane proteins, rotational and lateral diffusion resonance spectroscopy, *Biochim. Biophys. Acta*, 559, 289, 1979.
59. **Safman, P. G. and Delbruck, M.**, Brownian motion in biological membranes, *Proc. Natl. Acad. Sci. U.S.A.*, 72, 3111, 1975.
60. **Levine, Y. K., Birdsall, N. J. M., Lee, A. G., and Metcalfe, J. C.**, ^{13}C nuclear magnetic resonance relaxation measurements of synthetic lecithins and the effect of spin-labelled lipids, *Biochemistry*, 11, 1416, 1972.
61. **van Meer, G. and Op den Kamp, J. A. F.**, Transbilayer movement of various phosphatidylcholine species in intact human erythrocytes, *J. Cell. Biochem.*, 19, 193, 1982.
62. **Wirtz, K. W. A., Op den Kamp, J. A. F., and Roelofsen, B.**, Phosphatidylcholine transfer protein: properties and applications in membrane research, in *Progress in Protein-Lipid Interactions*, Watts, A. and De Pont, J. J. H. H. M., Eds., Elsevier, Amsterdam, 1986, chap. 7.
63. **Houslay, M. D. and Stanley, K. K.**, *Dynamics of Biological Membranes*, John Wiley & Sons, New York, 1982, chap. 2.
64. **Mohandas, N., Wyatt, J., Mel, S. F., Rossi, M. E., and Shohet, S. B.**, Phospholipid translocation across the human erythrocyte membrane, *J. Biol. Chem.*, 257, 6537, 1982.
65. **Zilversmit, D. B. and Hughes, M. E.**, Extensive exchange of rat liver microsomal lipids, *Biochim. Biophys. Acta*, 469, 99, 1977.
66. **van den Besselaar, A. M. H. P., de Kruijff, B., van den Bosch, H., and van Deenen, L. L. M.**, Phosphatidylcholine mobility in liver microsomal membranes, *Biochim. Biophys. Acta*, 510, 242, 1978.
67. **Jackson, R. L., Westerman, J., and Wirtz, K. W. A.**, Complete exchange of phospholipids between microsomes and plasma lipoproteins mediated by liver phospholipid-exchange proteins, *FEBS Lett.*, 94, 38, 1978.
68. **Gerritsen, W. J., de Kruijff, B., Verkleij, A. J., de Gier, J., and van Deenen, L. L. M.**, Ca^{2+}-induced isotropic motion and phosphatidylcholine flip-flop in phosphatidylcholine-cardiolipin bilayers, *Biochim. Biophys. Acta*, 598, 554, 1980.
69. **Noordam, P. C., van Echteld, C. J. A., de Kruijff, B., and de Gier, J.**, Rapid transbilayer movement of phosphatidylcholine in unsaturated phosphatidylethanolamine containing model membranes, *Biochim. Biophys. Acta*, 646, 483, 1981.
70. **Franck, P. F. H., Chiu, T. Y., Op den Kamp, J. A. F., Lubin, B., van Deenen, L. L. M., and Roelofsen, B.**, Accelerated transbilayer movement of phosphatidylcholine in sickled erythrocytes, *J. Biol. Chem.*, 258, 8435, 1983.
71. **Mohandas, C., Rossi, M., Bernstein, S., Ballas, S., Ravindranath, Y., Wyatt, J., and Mentzer, W.**, The structural organization of skeletal proteins influences lipid translocation across erythrocyte membrane, *J. Biol. Chem.*, 260, 14264, 1985.
72. **Kuypers, F. A., van Linde, M., Roelofsen, B., Tanner, M. J. A., Anstee, D. J., and Op den Kamp, J. A. F.**, Rh$_{null}$ human erythrocytes have an abnormal membrane phospholipid organization, *Biochem. J.*, 221, 931, 1984.
73. **Kuypers, F. A., van Linde, M., Roelofsen, B., Op den Kamp, J. A. F., Tanner, M. J. A., and Anstee, D. J.**, The phospholipid organization in the membrane of McLeod and Leach phenotype erythrocytes, *FEBS Lett.*, 184, 20, 1985.
74. **Poznansky, M. J. and Lange, Y.**, Transbilayer movement of cholesterol in dipalmitoyllecithin-cholesterol vesicles, *Nature*, 259, 420, 1976.
75. **Lenard, J. and Rothman, J. E.**, Transbilayer distribution and movement on cholesterol and phospholipid in the membrane of influenza virus, *Proc. Natl. Acad. Sci. U.S.A.*, 73, 391, 1976.
76. **Mabrey, S. and Sturtevant, J. M.**, Investigation of phase transitions of lipids and lipid mixtures by high sensitivity differential scanning calorimetry, *Proc. Natl. Acad. Sci. U.S.A.*, 73, 3862, 1976.
77. **Kim, J. T., Mattai, J., and Shipley, G. G.**, Bilayer interactions of ether- and ester-linked phospholipids: dihexadecyl- and dipalmitoylphosphatidylcholines, *Biochemistry*, 26, 6599, 1987.
78. **Sune, A. and Bienvenue, A.**, Relationship between transverse distribution of phospholipids in plasma membrane and shape change of human platelets, *Biochemistry*, 27, 6794, 1988.
79. **Ghosh, R.**, ^{31}P and ^{2}HNMR studies of structure and motion in bilayers of phosphatidylcholine and phosphatidylethanolamine, *Biochemistry*, 27, 7750, 1988.
80. **Casal, H. L. and Mantsch, H. H.**, Polymorphic phase behaviour of phospholipid membranes studied by infrared spectroscopy, *Biochim. Biophys. Acta*, 779, 381, 1984.
81. **Kornfeld, R. and Kornfeld, S.**, Assembly of asparagine-linked oligosaccharides, *Annu. Rev. Biochem.*, 54, 631, 1985.

82. **Goldfine, H., Johnston, N. C., Mattai, J., and Shipley, G. G.**, Regulation of bilayer in *Clostridium butyricum*: studies on the polymorphic phase behaviour of the ether lipids, *Biochemistry*, 26, 2814, 1987.
83. **Keough, K. M. W. and Taeusch, H. W., Jr.**, Surface balance and differential scanning calorimetric studies on aqueous dispersions of mixtures of dipalmitoyl phosphatidylcholine and short-chain, saturated phosphatidylcholines, *J. Colloid Interface Sci.*, 109, 364, 1986.
84. **Runquist, E. A. and Helmkamp, G. M.**, Design, synthesis, and characterization of bis-phosphatidylcholine, a mechanistic probe of phosphatidylcholine transfer protein catalytic activity, *Biochim. Biophys. Acta*, 940, 10, 1988.
85. **Harwood, J. L.**, Trans-bilayer lipid interactions, *Trends Biochem. Sci.*, 14, 2, 1989.
86. **Spiegel, S.**, Possible involvement of a GTP-binding protein in a late event during endogenous ganglioside-modulated cellular proliferation, *J. Biol. Chem.*, 264, 6766, 1989.
87. **Lennarz, W. J.**, Protein glycosylation in the endoplasmic reticulum: current topological issues, *Biochemistry*, 26, 7205, 1987.
88. **McIntosh, T. J., Magid, A. D., and Simon, S. A.**, Cholesterol modifies the short-range repulsive interactions between phosphatidylcholine membranes, *Biochemistry*, 28, 17, 1989.
89. **Murgolo, N. J., Patel, A., Stivale, S. S., and Wong, T. K.**, Dolichol acting as a membrane-fluidizing agent, *Biochemistry*, 28, 253, 1989.
90. **Gruner, S. M.**, Intrinsic curvature hypothesis for biomembrane lipid components: a role for non-bilayer lipids, *Proc. Natl. Acad. Sci. U.S.A.*, 82, 3665, 1985.
91. **Lewis, R. N. A. H., Mannock, D. A., McElhaney, R. N., Turner, D. C., and Gruner, S. M.**, Effect of fatty acyl chain length and structure on the lamellar gel to liquid-crystalline and lamellar to reversed hexagonal phase transitions of aqueous phosphatidylethanolamine dispersions, *Biochemistry*, 28, 541, 1989.
92. **Siegel, D. P., Banschbach, J., Alford, D., Ellens, H., Lis, L. J., Quinn, P. J., Yeagle, P. L., and Bentz, J.**, Physiological levels of diacylglycerol in phospholipid membranes induce membrane fusion and stabilize inverted phases, *Biochemistry*, 28, 3703, 1989.
93. **Banerjee, D. K., Kousvelari, E. E., and Baum, B. J.**, cAMP-mediated protein phosphorylation of microsomal membranes increases mannosylphosphodolichol synthase activity, *Proc. Natl. Acad. Sci. U.S.A.*, 84, 6389, 1987.

Chapter 4

ASYMMETRY IN MEMBRANE PHOSPHOLIPIDS

José M. Mato

TABLE OF CONTENTS

I. GENERAL COMMENTS

The asymmetric disposition of membrane proteins was the first to be observed. Proteins that span the membrane have a specific orientation. Thus, there is a part of the protein that is outside the cell which often contains carbohydrate residues; a small portion, containing mostly apolar amino acids, spans the phospholipid bilayer, and the remaining polar part is inside the cell. The asymmetry of membrane proteins is therefore absolute, indicating that protein transbilayer movements does not occur.[1]

Research on phospholipid asymmetry was initiated after observing that some of the labeling reagents used to explore protein asymmetry also reacted with certain phospholipids. These reagents are used for specific labeling of one-half of the bilayer in sealed membrane vesicles or intact cells. The pattern of labeling under such conditions is then compared with that observed with preparations of broken cells or inverted sealed membranes (Figure 1). To be used to explore membrane asymmetry, these reagents must not be able to enter the sealed vesicle nor induce alterations in the original phospholipid distribution. The first reagents used to label phospholipids were 4-acetamido-4′-isothiocyano-2,2′-stilbene disulphonate (SITS), which is fluorescent, and formylmethionyl sulphone methylphosphate (FMMP) (Figure 2).[2,3] An interesting pair of reagents to study lipid asymmetry in membranes are isethionyl acetimidate (IAI) and ethyl acetimidate (EAI) (Figure 2).[4] Whereas IAI cannot penetrate the bilayer and thus labels only the cell surface, EAI enters the cell and labels both sides.[4] The reagents mentioned above react with amino phospholipids and have been successfully used to determine the asymmetric distribution of phosphatidylethanolamine and phosphatidylserine in various cellular membranes.

II. PHOSPHOLIPID ASYMMETRY IN THE MAMMALIAN ERYTHROCYTE MEMBRANE

The idea that phospholipids were asymmetrically distributed in erythrocyte cell membranes came from the observation that relatively little phosphatidylethanolamine or phosphatidylserine reacted with SITS or FMMP, which reacted with the free amino groups of phosphatidylethanolamine and phosphatidylserine, but not with the trimethyl amino group of phosphatidylcholine, versus the labeling obtained using unsealed cell ghosts, where both sides of the bilayer are accessible to the reagents.[2,3] Treatment with both IAI and EAI also led to the conclusion that, in general, phosphatidylethanolamine and phosphatidylserine are preferentially localized in the inner surface of the cell membrane.[4] IAI has also been used to localize, at the cell surface of rat hepatocytes, a glycosyl phosphatidylinositol which might be involved in insulin action (see Chapter 8). This glycolipid contains a residue of non-N-acetylated glucosamine which reacted with [^{14}C]-IAI forming a radiolabeled glycolipid derivative.[5]

Another approach to the study of phospholipid asymmetry has been the use of specific lipases. Lipases are used to hydrolyze specific phospholipids at one side of the membrane using sealed vesicles or intact cells, and the results are compared with those obtained using broken cells or inverted vesicles. The use of enzymes to study lipid asymmetry has several drawbacks. First, the formation of diacylglycerol or lysophospholipids by the action of phospholipase C or phospholipase A$_2$, respectively, might perturb the structure of the membrane (see Chapter 3). Thus, diacylglycerol tends to aggregate into microdomains[6-9] which, together with the loss of polar headgroups, might induce flip-flop of phospholipids.[10] Second, several phospholipases have a limited capacity to penetrate the membrane and do not have access to phospholipids in the intact erythrocyte, whereas they rapidly hydrolyze phospholipids in a broken erythrocyte.[11-13] Despite these difficulties, the available results indicate that the majority (about 85%) of the sphingomyelin could be hydrolyzed by sphingomyelinase treatment of intact erythrocytes, whereas phosphatidylserine, phosphatidylinositol, and phos-

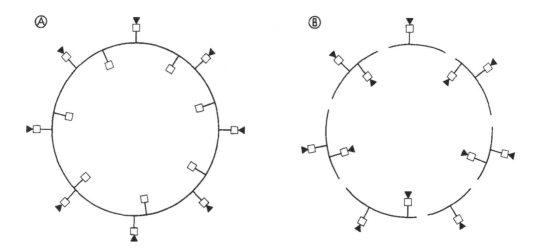

FIGURE 1.   In A (intact cells), the reagent (▼) reacts with the molecules of phospholipid (□) that are present at the outer surface of the cell. In B (broken cells), the reagent (▼) reacts with the molecules of phospholipid that are at both sides of the cell. By comparing the labeling in A and B, the distribution of the phospholipid (□) between both surfaces can be determined.

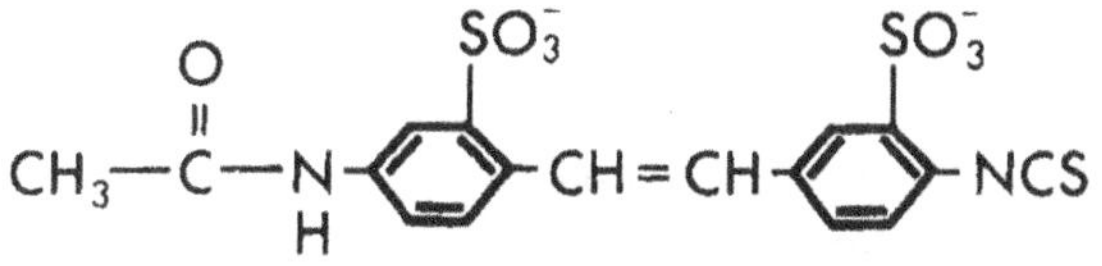

$$CH_3-\overset{O}{\underset{}{C}}-\underset{H}{N}H-\underset{}{\bigcirc}^{SO_3^-}-CH=CH-\bigcirc^{SO_3^-}-NCS \qquad SITS$$

4—acetimido—4'—isothiocyano—2,2'—stilbene disulphonate

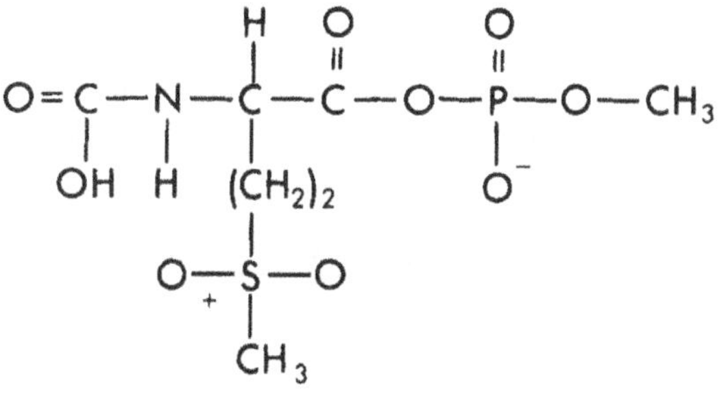

$$O=\underset{OH}{C}-\underset{H}{N}-\underset{(CH_2)_2}{\overset{H}{C}}-\overset{O}{\underset{}{C}}-O-\overset{O}{\underset{O^-}{P}}-O-CH_3 \qquad FMMP$$

$$O-\overset{}{\underset{+}{S}}-O$$
$$CH_3$$

Formylmethionylsulphone methylphosphate

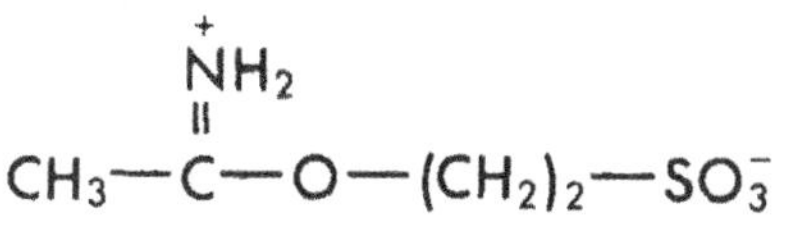

$$CH_3-\overset{\overset{+}{N}H_2}{\underset{}{C}}-O-(CH_2)_2-SO_3^- \qquad IAI$$

Isethionylacetimidate

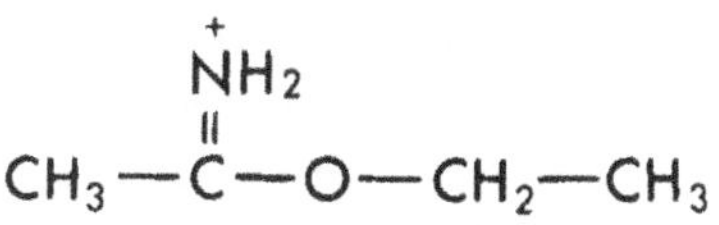

$$CH_3-\overset{\overset{+}{N}H_2}{\underset{}{C}}-O-CH_2-CH_3 \qquad EAI$$

Ethyl acetimidate

FIGURE 2.   Structure of several reagents commonly used to determine the asymmetric distribution of phospholipids in biomembranes.

phatidylethanolamine, but not phosphatidylcholine, are resistant to phospholipase A$_2$ (or phospholipase C) treatment, which readily degrades all four phospholipids when added to unsealed cell ghosts.[12-17]

A different approach to studying the asymmetric distribution of phospholipids has been the use of phospholipid exchange proteins (see Chapter 3). By incubating intact cells or sealed membrane vesicles with liposomes containing a given radioactive phospholipid (phosphatidylcholine, phosphatidylserine, phosphatidylethanolamine, or sphingomyelin) in the presence of the specific exchange protein, it is possible to measure the size of the exchangeable pool of this lipid. This represents the fraction of the phospholipid which is present at the external surface of the cell or lipid vesicle. The results from this type of experimental design also indicate that the majority of phosphatidylcholine and sphingomyelin are at the cell surface, whereas phosphatidylethanolamine, phosphatidylserine, and phosphatidylinositol are mainly localized at the inner side of the cellular bilayer.[17-19]

Glycosphingolipids are also confined to the external surface of the cell.[20,21] To reach this conclusion, erythrocytes were specifically labeled in galactose and galactosamine residues by treatment with [^{3}H]-borohydride after exposure of the cell to galactose oxidase. Whereas [^{3}H]-borohydride might cross the bilayer, galactose oxidase would not, and the topographical distribution of glycosphingolipids could thus be investigated. The pattern of glycosphingolipid labeling is the same for intact erythrocytes and unsealed ghosts, indicating that the majority of the glycolipids are present at the cell surface. By using this technique, two glycosyl phosphatidylinositols have been recently localized at the outer surface of *Leishmania major* (also see Chapter 5).[69]

The topographical distribution of cellular cholesterol has also been investigated by determining the exchange of [^{3}H]-cholesterol between phospholipid:cholesterol vesicles and an excess of erythrocyte ghosts,[22] and by treatment of intact cells with cholesterol oxidase.[61] These experiments indicate that about 70% of the cholesterol is at the outer surface of the erythrocyte.

In conclusion, the available evidence indicates that in human erythrocytes over 75% of the phosphatidylcholine, about 85% of the sphingomyelin, 70% of the cholesterol, and all of the glycosphingolipids are at the outer surface of the cell, whereas the inner surface contains 80% of the phosphatidylethanolamine and all of the phosphatidylserine and phosphatidylinositol.

III. ABNORMAL PHOSPHOLIPID ASYMMETRY IN SICKLE ERYTHROCYTES

The primary defect in sickle cell anemia is one involving polymerization of sickle hemoglobin. Secondary to this hemoglobin defect, a number of membrane abnormalities have been detected in sickle erythrocytes. These include an increased cation leak and calcium accumulation,[24-26] increased hemoglobin binding to the cell membrane,[27] abnormal protein phosphorylation,[28] and increased lipid peroxidation.[29] Abnormalities in the accessibility of phospholipids to a variety of exogenous reagents (dinitrofluorobenzene [DNFB], trinitrobenzene sulfonic acid [TNBS], and phospholipase A$_2$) have also been reported in sickle erythrocytes. Thus, the accessibility of phosphatidylethanolamine and phosphatidylserine to DNFB and TNBS has been found to be markedly decreased in both irreversible sickle cells and deoxygenated reversible sickle cells, compared to both normal and oxygenated reversible sickle cells.[30,31] Similarly, the degradation of phosphatidylethanolamine and phosphatidylserine by phospholipase A$_2$ was increased in deoxygenated reversible sickle cells, compared to both normal and oxygenated reversible sickle cells.[32] The accessibility of sphingomyelin in sickle cells, however, was similar to that found in normal erythrocytes.[32] Interestingly, reoxygenation of reversible sickle cells, but not of irreversible sickle cells, almost completely restored the normal accessibility of phospholipids by phospholipase A$_2$ and, therefore, the distribution of phospholipids between both sides of the membrane.[32] Using a specific phosphatidylcholine transfer protein, an accelerated transbilayer movement of phosphatidylcholine in sickle erythrocytes has been shown.[33] In deoxygenated reversible sickle cells, all of the phosphatidylcholine present in the membrane of the intact cell is rapidly available for exchange, mediated by the phosphatidylcholine transfer protein. Since a substantial amount of this phospholipid is present in the inner surface of the membrane, this observation indicates that phosphatidylcholine in these cell membranes undergoes fast transbilayer, flip-flop movements. This enhanced transbilayer movement of phosphatidylcholine, detected in deoxygenated reversible sickle cells, is rapidly restored to the normal slow rate upon reoxygenation of the cells, indicating that the process is reversible. The half-time for phosphatidylcholine transbilayer movement in reversible sickle cells is around 3.5 h,[33] whereas the half-time in normal cells is between 13 and 27 h (see Chapter 3). It is important to note that the half-time for translocation in reversible sickle cells, although faster than in normal cells, is still much slower than the diffusion rates of phospholipids in the monolayer plane (see Chapter 3). However, it might be important, compared to the average life span of the erythrocyte (120 d in humans).

This situation in sickle red cells is comparable to that induced by oxidative cross-linking of membrane proteins of normal erythrocytes, where amino phospholipids have also been reported to be more accessible to phospholipase A$_2$ than in normal cells.[34] These results suggest that the asymmetric distribution of phospholipids is closely related to the structural integrity of the membrane proteins, which are known to posses an absolute asymmetry. Since in sickle cells there is a major alteration of the spectrin-actin lattice,[35] it has been suggested that the structural integrity of the erythrocyte cytoskeleton might be essential to maintain the correct organization of the lipid bilayer.[36,37] However, it has recently been reported that spectrin and other cytoskeleton proteins are not major factors in the establishment and maintenance of phospholipid asymmetry in human erythrocytes, which may be mainly due to the aminophospholipid translocase activity.[65]

IV. PHOSPHOLIPID ASYMMETRY IN THE PLASMA MEMBRANE OF CELLS DIFFERENT FROM ERYTHROCYTES

Phospholipid asymmetry has been investigated in plasma membranes from a variety of tissues, including rat liver, pig platelet, mouse LM cells, and hamster kidney RHK-21 cells.[38-43] Despite large differences in the phospholipid composition of plasma membranes from these cells, they all show the same basic pattern of phospholipid asymmetry, with phosphatidylethanolamine, phosphatidylserine, and phosphatidylinositol facing preferentially the cytoplasmic side of the membrane and phosphatidylcholine, sphingomyelin, and glycolipids on the cell surface.

V. PHOSPHOLIPID ASYMMETRY IN SUBCELLULAR MEMBRANES

The asymmetric distribution of phospholipids in the inner membrane of beef heart mitochondria has been investigated by chemical labeling with nonpenetrating probes, such as isethionyl acetimidate, fluorescamine, and trinitrobenzenesulfonate, phospholipase A$_2$ treatment for the selective degradation of exposed phospholipids, and reaction with anticardiolipin antibodies.[44-47] In these experiments, right-side-out and inside-out mitochondrial inner membranes have been used. The results from these experiments indicate that phospholipids of the inner mitochondrial membrane are oriented according to their charge. Negatively charged cardiolipin and phosphatidylinositol are oriented preferentially on the matrix side of the inner mitochondrial membrane (about 80%) and phosphatidylcholine and phosphatidylethanolamine show a certain preference for the cytoplasmic side (60%).[48] Phospholipid asymmetry has been also reported in microsome, Golgi, and lysosomal membranes.[45,49,50] These results indicate that the asymmetric distribution of phospholipids is probably a general feature of all biological membranes.

VI. ROLE OF PHOSPHOLIPID ASYMMETRY IN MEMBRANE FUNCTION

There is clearly no randomized distribution of phospholipids between both layers of a biomembrane. The factors determining phospholipid asymmetry are not well understood, but the existence of an outward as well as an inward motion has been proposed. Phospholipid asymmetry is probably related to the topology of membrane proteins and to the structural integrity of the cytoskeleton.[36,37] The asymmetric distribution of phospholipids is specific to a particular cellular membrane. This suggests that the phospholipids of each leaflet of the membrane might have specific cellular functions. The asymmetric distribution of phospholipid headgroups might generate two layers that differ in fluidity (e.g., there is an apparent enrichment of unsaturated fatty acids on the cytoplasmic side of the plasma membrane in LM cells[40]) and charge, which might be important in protein-lipid interactions.

Phosphatidylserine and phosphatidylethanolamine are more reactive, and the predominant distribution of these phospholipids on the cytoplasmic side of the cell might prevent them from reacting with the extracellular environment. Thus, the abnormal presence of phosphatidylserine on the outer surface of the membrane has major pathophysiological consequences. The translocation to the outer cell surface of phosphatidylserine in sickle erythrocytes can promote clot formation, and this abnormality might contribute to the pathogenesis of the vaso-occlusive episodes in sickle cell anemia.[51] The mechanism responsible for transbilayer movement of phosphatidylserine is not well understood. Exogenously added phosphatidylserine translocates from the outer surface, where insertion takes place, to the inner membrane leaflet with a half-time of less than 10 min.[52] A variety of derivatives of pyridyldithioethylamine have been synthesized which specifically and reversibly inhibit transbilayer movement of phosphatidylserine.[53] A ^{125}I-labeled derivative of pyridylthioethylamine has been shown to react preferentially with a 31-kDa protein which might be involved in phosphatidylserine transport.[54] Similarly, labeling of human erythrocyte membrane proteins by photoactivable radioiodinated phosphatidylserine revealed seven membrane proteins with an M_r between 140 and 27 kDa;[55] whether any of these proteins are involved in phosphatidylserine translocation remains to be determined. In chromaffin granules, there is also evidence for an ATP-dependent aminophospholipid translocase.[68] Stimulation of platelets with the Ca^{2+} ionophore A23187 or by the combined action of collagen plus thrombin results in a rapid loss of the asymmetric distribution of phosphatidylserine.[62] Moreover, treatment with sulfhydryl reagents caused the exposure of phosphatidylserine at the outer surface, and endogenous phosphatidylserine, previously exposed at the outer surface during cell activation or sulfhydryl group modification, can be translocated back to the cytoplasmic leaflet of the membrane by addition of dithiothreitol.[62,66] These results have been interpreted as indicative of the existence of an aminophospholipid-specific translocase in the platelet membrane.[62,66] The ratio of cholesterol to phospholipid may play a role in controlling aminophospholipid translocase activity. Thus, cholesterol depletion accelerates the outside/inside passage of aminophospholipids, while cholesterol enrichment has the opposite effect.[66] Transmembrane lipid asymmetry has been observed to change during myoblast differentiation, although the mechanisms responsible for these changes are unknown.[63] During adipocyte differentiation of 3T3F442A cells, there is a decrease in plasma lipid order which is probably due to an increase in the amount of monounsaturated phospholipid acyl chains and a decrease in the ratio of cholesterol to phospholipid.[64] Whether these changes can influence membrane asymmetry as well as the physical properties of the bilayer is not known.

The outside → inside and inside → outside translocation of phospholipids has been studied in a variety of cells, using spin-labeled phospholipids. Inward transport of aminophospholipids is much faster than that of phosphatidylcholine and is mediated by a specific ATP-dependent protein,[56-59] and perhaps by a pH gradient.[67] The rate of outward motion of aminophospholipids, which is protein mediated, is also faster than that of phosphatidylcholine, but the differences are smaller between the outward than between the inward translocation rates.[60]

The predominant localization of phosphatidylinositol and its phosphorylated derivatives in the inner surface of the cell is important for the generation of inositol phosphates and diacylglycerol in the cytosolic side of the bilayer, where they serve as mediators of the action of growth factors and hormones (also see Chapter 7).[70] Glycolipids, which appear to be localized almost exclusively at the outer surface of the cell, have been implicated in a variety of processes involving cellular interactions, differentiation, and oncogenesis (also see Chapter 9).[53]

REFERENCES

1. **Steck, T.**, The organization of proteins in the human red blood cell membrane, *J. Cell. Biol.*, 62, 1, 1974.
2. **Maddy, A. H.**, A fluorescent label for the outer components of the plasma membrane, *Biochim. Biophys. Acta*, 88, 390, 1964.
3. **Bretscher, M. S.**, Asymmetrical lipid bilayer structure of biological membranes, *Nature New Biol.*, 236, 11, 1972.
4. **Whiteley, N. M. and Berg, H. C.**, Amidination of the outer and inner surfaces of the human erythrocyte membrane, *J. Mol. Biol.*, 87, 541, 1974.
5. **Alvarez, J. F., Varela, I., Ruiz-Albusac, J. M., and Mato, J. M.**, Localisation of the insulin-sensitive phosphatidylinositol glycan at the outer surface of the cell membrane, *Biochem. Biophys. Res. Commun.*, 152, 1455, 1988.
6. **Coleman, R., Finean, J. B., Knutton, S., and Limbrick, A. R.**, A structural study of the modification of erythrocyte ghosts by phospholipase C, *Biochim. Biophys. Acta*, 219, 81, 1970.
7. **van Deenen, L. L. M.**, The nonrandom nature of membrane organization, in *Functional Linkage in Biomolecular Systems*, Schmitt, F. O., Schneider, D. M., and Crothers, D. M., Eds., Raven Press, New York, 1975, 106.
8. **Verkleij, A. J., Zwaal, R. F. A., Roelofsen, B., and van Deenen, L. L. M.**, The asymmetric distribution of phospholipids in the human red cell membrane. A combined study using phospholipases and freeze-etch electron microscopy, *Biochim. Biophys. Acta*, 323, 178, 1973.
9. **Mountford, C. E. and Wright, L. C.**, Organization of lipids in the plasma membrane of malignant and stimulated cells: a new model, *Trends Biochem. Sci.*, 13, 172, 1988.
10. **Houslay, M. D. and Stanley, K. K.**, *Dynamics of Biological Membranes*, John Wiley & Sons, New York, 1982, chap. 4.
11. **Zwaal, R. F. A., Roelofsen, B., and Colley, C. M.**, Localization of red cell membrane constituents, *Biochim. Biophys. Acta*, 300, 159, 1973.
12. **Roelofsen, B., Zwaal, R. F. A., Comfurius, P., Woodward, C. B., and van Deenen, L. L. M.**, Action of pure phospholipase A_2 and phospholipase D on human erythrocytes and ghosts, *Biochim. Biophys. Acta*, 241, 925, 1971.
13. **Ibrahim, S. A. and Thompson, R. H. S.**, Action of phospholipase A on human red cell ghosts and intact erythrocytes, *Biochim. Biophys. Acta*, 99, 331, 1964.
14. **Demel, R. A., Geurtz van Kessel, W. S. M., Zwaal, R. F. A., Roelofsen, B., and van Deenen, L. L. M.**, Relation between various phospholipase actions on human red cell membranes and the interfacial phospholipid pressure in monolayers, *Biochim. Biophys. Acta*, 406, 97, 1975.
15. **Verkleij, A. J., Zwaal, R. F. A., Roelofsen, B., Comfurius, P., Kastelijn, D., and van Deenen, L. L. M.**, The asymmetric distribution of phospholipids in the human red cell membrane. A combined study using phospholipases and freeze-etch electron microscopy, *Biochim. Biophys. Acta*, 323, 178, 1973.
16. **Renooij, W., van Golde, L. M. G., Zwaal, R. F. A., and van Deenen, L. L. M.**, Topological asymmetry of phospholipid metabolism in rat erythrocyte membranes, *Eur. J. Biochem.*, 61, 53, 1976.

17. **Gazitt, Y., Loyter, A., Reichler, Y., and Ohad, I.,** Correlation between changes in the membrane organization and susceptibility to phospholipase C attack induced by ATP depletion of rat erythrocytes, *Biochim. Biophys. Acta,* 419, 479, 1976.

18. **Wirtz, K. W. A.,** Transfer of phospholipids between membranes, *Biochim. Biophys. Acta,* 344, 95, 1974.

19. **Op den Kemp, J. A. F.,** Lipid asymmetry in membranes, *Annu. Rev. Biochem.,* 48, 47, 1979.

20. **van Meer, G., Poorthuis, B. J. H. M., Wirtz, K. W. A., Op den Kamp, J. A. F., and van Deenen, L. M.,** The transbilayer distribution and mobility of phosphatidylcholine in intact erythrocyte membranes, *Eur. J. Biochem.,* 103, 283, 1980.

21. **Gahmberg, C. G. and Hakomori, S. I.,** External labeling of cell surface galactose and galactosamine in glycolipid and glycoprotein of human erythrocytes, *J. Biol. Chem.,* 248, 4311, 1973.

22. **Steck, T. L. and Dawson, G.,** Topographical distribution of complex carbohydrates in the erythrocyte membrane, *J. Biol. Chem.,* 249, 2135, 1974.

23. **Poznansky, M. J. and Lange, Y.,** Transbilayer movement of cholesterol in phospholipid vesicles under equilibrium and non-equilibrium, *Biochim. Biophys. Acta,* 506, 256, 1978.

24. **Tosteson, D. C., Carlson, E., and Dunham, E. T.,** The effect of sickling on ion transport. I. The effect of sickling on potassium transport, *J. Gen. Physiol.,* 39, 31, 1955.

25. **Glader, B. E., Muller, A., and Nathan, D. G.,** Comparison of membrane permeability abnormalities in reversibly sickled erythrocytes, in Proc. 1st Natl. Symp. Sickle Cell Disease, Hercules, J. I., Schnechter, A. N., Eaton, W. A., and Jackson, R. E., Eds., Department of Health, Education, and Welfare, 1974, 75.

26. **Eaton, J. W., Skelton, T. D., Swofford, H. S., Kolpin, C. E., and Jacob, H. S.,** Elevated erythrocyte calcium in sickle cell disease, *Nature,* 246, 105, 1973.

27. **Asakura, T. K., Minakaza, K., Adachi, K., Russel, M. O., and Schwartz, E.,** Denatured hemoglobin in sickle erythrocytes, *J. Clin. Invest.,* 59, 633, 1976.

28. **Hosey, M. M. and Tao, M.,** Altered erythrocyte membrane phosphorylation in sickle cell disease, *Nature,* 263, 424, 1976.

29. **Lubin, B. and Chiu, D.,** Abnormal susceptibility of sickle erythrocytes to lipid peroxidation, in *Red Blood Cell and Lens Metabolism,* Srivastava, S., Ed., Elsevier, New York, 1980, 159.

30. **Gordesky, S. E., Marinetti, G. V., and Segel, B. G.,** Differences in the reactivity of phospholipids with FDNB in normal RBC, sickle cells and RBC ghosts, *Biochem. Biophys. Res. Commun.,* 47, 1004, 1972.

31. **Chiu, D., Lubin, B., and Shohet, S.,** Erythrocyte membrane lipid reorganization during the sickling process, *Br. J. Haematol.,* 41, 223, 1979.

32. **Lubin, B., Chiu, D., Bastacky, J., Roelofsen, B., and van Deenen, L. L. M.,** Abnormalities in membrane phospholipid organization in sickled erythrocytes, *J. Clin. Invest.,* 67, 1643, 1981.

33. **Franck, P. F. H., Chiu, D., Op den Kamp, J. A. F., Lubin, B., van Deenen, L. L. M., and Roelofsen, B.,** Accelerated transbilayer movement of phosphatidylcholine in sickled erythrocytes, *J. Biol. Chem.,* 258, 8435, 1983.

34. **Haest, C. W. M. and Deuticke, B.,** Possible relationship between membrane proteins and phospholipid asymmetry in the human erythrocyte membrane, *Biochim. Biophys. Acta,* 436, 353, 1976.

35. **Lux, S. E., John, K. M., and Karnovsky, M. J.,** Irreversible deformation of the spectrin-actin lattice in irreversibly sickled cells, *J. Clin. Invest.,* 58, 955, 1976.

36. **Haest, C. W. M., Kamp, D., and Deuticke, B.,** Spectrin as a stabilizer of the phospholipid asymmetry in the human erythrocyte membrane, *Biochim. Biophys. Acta,* 509, 21, 1978.

37. **Franck, P. F. H., Roelofsen, B., and Op den Kamp, J. A. F.,** Complete exchange of phosphatidylcholine from intact erythrocytes after protein crosslinking, *Biochim. Biophys. Acta,* 687, 105, 1982.

38. **Higgins, J. A. and Evans, W. H.,** Transverse organization of phospholipid across the bilayer of plasma-membrane subfractions of rat hepatocytes, *Biochem. J.,* 174, 563, 1978.

39. **Zwaal, R. F. A.,** Membrane and lipid involvement in blood coagulation, *Biochim. Biophys. Acta,* 515, 163, 1978.

40. **Sandra, A. and Pagano, R. E.,** Phospholipid asymmetry in LM cell plasma membrane derivatives: polar head group and acyl chain distributions, *Biochemistry,* 17, 332, 1978.

41. **Patzer, E. J., Moore, N. F., Barenholz, Y., Shaw, J. M., and Wagner, R. R.,** Lipid organization of the membrane of vesicular stomatitis virus, *J. Biol. Chem.,* 253, 4544, 1978.

42. **Fontaine, R. N. and Schroeder, F.,** Plasma membrane aminophospholipid distribution in transformed murine fibroblasts, *Biochim. Biophys. Acta,* 558, 1, 1979.

43. **Higgins, J. A. and Evans, W. H.,** Transverse organization of phospholipid across the bilayer of plasma-membrane subfractions of rat hepatocytes, *Biochem. J.,* 174, 563, 1978.

44. **Krebs, J. J. R., Hauser, H., and Carafoli, E.,** Asymmetric distribution of phospholipids in the inner membrane of beef heart mitochondria, *J. Biol. Chem.,* 254, 5308, 1979.

45. **Nilsson, O. S. and Dallner, G.,** Transverse asymmetry of phospholipids in subcellular membranes of rat liver, *Biochim. Biophys. Acta,* 464, 453, 1977.

46. **Crain, R. C. and Marinetti, G. V.**, Phospholipid topology of the inner mitochondrial membrane of rat liver, *Biochemistry*, 18, 2407, 1979.

47. **Harb, J. S., Comte, J., and Gautheron, D. C.**, Asymmetrical orientation of phospholipids and their interactions with marker enzymes in pig heart mitochondrial inner membrane, *Arch. Biochem. Biophys.*, 208, 305, 1981.

48. **Daum, G.**, Lipids of mitochondria, *Biochim. Biophys. Acta*, 822, 1, 1985.

49. **Higgins, J. A. and Dawson, R. M. C.**, Asymmetry of the phospholipid bilayer of rat liver endoplasmic reticulum, *Biochim. Biophys. Acta*, 470, 342, 1977.

50. **Higgins, J. A.**, The transverse distribution of phospholipids in the membranes of Golgi subfractions of rat hepatocytes, *Biochem. J.*, 219, 262, 1984.

51. **Chiu, D., Lubin, B., Roelofsen, B., and van Deenen, J. L. M.**, Sickled erythrocytes accelerated clotting in vitro: an effect of abnormal membrane lipid asymmetry, *Blood*, 58, 398, 1981.

52. **Sune, A. and Bienvenue, A.**, Relationship between transverse distribution of phospholipids in plasma membrane and shape change of human platelets, *Biochemistry*, 27, 6794, 1988.

53. **Connor, J. and Schroit, A. J.**, Transbilayer movement of phosphatidylserine in erythrocytes: inhibition of transport and preferential labeling of a 31,000-dalton protein by sulfhydryl reactive reagents, *Biochemistry*, 27, 848, 1988.

54. **Feizi, T.**, Demonstration by monoclonal antibodies that carbohydrate structures of glycoproteins and glycolipids are onco-developmental antigens, *Nature*, 314, 53, 1985.

55. **Zachowski, A., Fellmann, P., Hervé, P., and Devaux, P. F.**, Labeling of human erythrocyte membrane proteins by photoactivable radiodinated phosphatidylcholine and phosphatidylserine, *FEBS Lett.*, 223, 315, 1987.

56. **Seigneuret, M. and Devaux, P. F.**, ATP-dependent asymmetric distribution of spin-labeled phospholipids in the erythrocyte membrane: relation to shape changes, *Proc. Natl. Acad. Sci. U.S.A.*, 81, 3751, 1984.

57. **Daleke, D. L. and Huestis, W. H.**, Incorporation and translocation of aminophospholipids in human erythrocytes, *Biochemistry*, 24, 5406, 1985.

58. **Zachowski, A., Favre, E., Cribier, S., Herve, P., and Devaux, P. F.**, Outside-inside translocation of aminophospholipids in the human erythrocyte membrane is mediated by a specific enzyme, *Biochemistry*, 25, 2585, 1986.

59. **Martin, O. C. and Pagano, R. E.**, Transbilayer movement of fluorescent analogs of phosphatidylserine and phosphatidylethanolamine at the plasma membrane of cultured cells, *J. Biol. Chem.*, 262, 5890, 1987.

60. **Bitbol, M. and Devaux, P. F.**, Measurement of outward translocation of phospholipids across human erythrocyte membrane, *Proc. Natl. Acad. Sci. U.S.A.*, 85, 6783, 1988.

61. **Lange, Y. and Ramos, B. V.**, Analysis of the distribution of cholesterol in the intact cell, *J. Biol. Chem.*, 258, 15130, 1983.

62. **Bevers, E. M., Tilly, R. H. J., Senden, J. M. G., Comfurius, P., and Zwaal, R. F. A.**, Exposure of endogenous phosphatidylserine at the outer surface of stimulated platelets is reversed by restoration of aminophospholipid translocase activity, *Biochemistry*, 28, 2382, 1989.

63. **Sessions, A. and Horowitz, A. F.**, Differentiation-related differences in the plasma membrane phospholipid asymmetry of mitogenic fibrogenic cells, *Biochim. Biophys. Acta*, 728, 103, 1983.

64. **Storch, J., Shulman, S. L., and Kleinfeld, A. M.**, Plasma membrane lipid order and composition during adipocyte differentiation of 3T3F442A cells, *J. Biol. Chem.*, 264, 10527, 1989.

65. **Calvez, J. Y., Zachowski, A., Herrmann, A., Marrot, G., and Devaux, P. F.**, Asymmetric distribution of phospholipids in spectrin-poor erythrocyte vesicles, *Biochemistry*, 27, 5666, 1989.

66. **Marrot, G., Herve, P., Zachowski, A., Fellmann, P., and Devaux, P. F.**, Aminophospholipid translocase of human erythrocytes: phospholipid substrate specificity and effects of cholesterol, *Biochemistry*, 28, 3456, 1989.

67. **Hope, M. J., Redelmeier, T. E., Wong, K. F., Rodriguez, W., and Cullis, P. R.**, Phospholipid asymmetry in large unilamellar vesicles induced by transmembrane pH gradients, *Biochemistry*, 28, 4181, 1989.

68. **Zachowski, A., Henry, J. P., and Devaux, P. F.**, Control of transmembrane lipid asymmetry in chromaffin granules by an ATP-dependent protein, *Nature*, 340, 75, 1989.

69. **Rosen, G., Pahlsson, P., Londner, M. V., Westerman, M. E., and Nilsson, B.**, Structural analysis of glycosyl-phosphatidylinositol antigens of *Leishmania major*, *J. Biol. Chem.*, 264, 10457, 1989.

70. **Berridge, M. B.**, Inositol triphosphate and diacylglycerol: two interacting second messengers, *Annu. Rev. Biochem.*, 56, 159, 1987.

Chapter 5

ROLE OF PHOSPHATIDYLINOSITOL IN PROTEIN ATTACHMENT

José M. Mato and Isabel Varela

TABLE OF CONTENTS

I. GENERAL COMMENTS

Biomembranes are mixtures of lipids and proteins. Membrane proteins are arbitrary classified in two groups: peripheral and integral proteins. Proteins which might be removed by washing the membrane with low or high ionic strength solutions are termed peripheral proteins. Integral proteins are defined as those that require detergents for extraction. Peripheral proteins are bound to the polar regions of other proteins or lipids. In most models for biological membranes, integral proteins cross the lipid bilayer, via a hydrophobic protein sequence, exposing two hydrophylic moieties, one at the outer surface and the other at the cytoplasmic side of the cell (see Figure 1). Proteins can either cross the membrane once (as is normally the case with growth factor receptors) or several times (as is normally the case with transport proteins).

Integral proteins undergo extensive hydrophobic and ionic interactions with the surrounding lipids. In fact, proteins act as impurities, perturbing the structure of the phospholipid bilayer. In general, proteins are immersed in the liquid-crystalline regions of the bilayer, being excluded from the less flexible solid-state structures.[1] In addition, integral proteins might be anchored to the membrane exclusively by a covalently attached phospholipid. An increasing number of proteins are now known to be anchored to the membrane via the carboxy terminus, which is attached to the ethanolamine moiety of a glycosyl phosphatidylinositol (glycosyl-PI) molecule (Figure 1). Table 1 lists those proteins which are believed to be attached to the membrane exclusively by a glycosyl-PI anchor.

II. INOSITOL CONTAINING GLYCOPHOSPHOLIPIDS AS PROTEIN ANCHORS

Inositol glycophospholipids with the general structure lipid-phosphate-inositol-glycan have been isolated from plants,[2] bacteria,[3] yeast,[4] protozoa,[5,6] and mammalian cells.[7-9] Some of these glycosyl-PIs serve in protozoa as well as in mammalian organisms to attach proteins to the cell surface,[10-13] others give rise to a phospho-oligosaccharide, which might act as a mediator of some of the actions of insulin (also see Chapter 8),[12,14] and the function of others remain to be determined. Preliminary evidence about glycosyl-PI acting as protein anchors was given in the early 1960s,[15,16] but it is only recently that we have begun to understand the structure of this anchor and its importance in maintaining cell function and structure.[10-14] Table 1 lists proteins which are believed to be attached to the membrane exclusively by a glycosyl-PI. All the proteins listed in Table 1 are present at the cell surface, with the sole exception of ornitine decarboxilase and myelin basic protein, which are believed to be present in the inner surface of mouse lymphocytes and bovine brain white matter, respectively, through a glycosyl-PI,[17,18] and two membrane proteins with M_rs of 82 and 68 kDa, respectively, that have been reported to be attached to granule membranes of chromaffin cells through glycosyl-PI anchors.[102] Although the presence of *myo*-inositol has been documented in ornitine decarboxilase, there is no direct evidence indicating that this enzyme is glycosyl-PI anchored. The enzyme was activated by treatment with a phosphatidylinositol-specific phospholipase C (PI-PlC), but there are no data about the release of ornitine decarboxilase by PI-PlC incubation.[17] In the case of myelin basic protein, the attached phospholipid has been identified as phosphatidylinositol-4-phosphate and/or phosphatidylinositol-4,5-bisphosphate.[18] As with ornitine decarboxylase, there is no direct evidence indicating that phosphatidylinositol is involved in attaching the myelin basic protein to the membrane.[18] Therefore, although listed in Table 1, further evidence is required to definitively include ornitine decarboxilase and/or myelin basic protein in the family of glycosyl-PI-anchored proteins.

As mentioned above, two membrane proteins with M_rs of 82 and 68 kDa, respectively, have been reported to be attached to granule membranes of chromaffin cells through glycosyl-

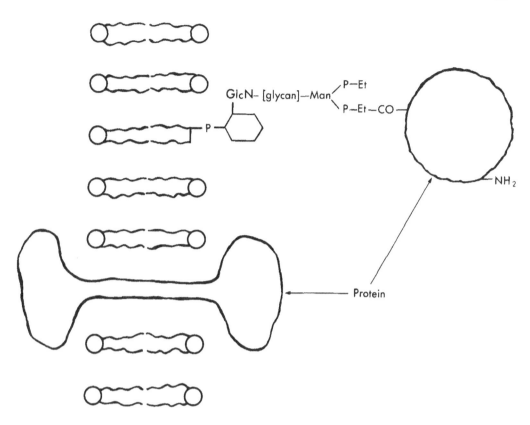

FIGURE 1. Schematic representation of biological membranes showing an integral protein crossing the lipid bilayer via a hydrophobic protein sequence and exposing two hydrophylic moieties, one at the outer surface and the other at the cytoplasmic side of the cell. The figure also shows an integral protein anchored to the membrane exclusively via the carboxy terminus, which is attached to the ethanolamine moiety of a glycosyl phosphatidylinositol molecule.

PI anchors.[102] The evidence in this case included solubilization by treatment of the membranes with PI-PlC and nitrous acid deamination. Using polarized epithelial cell monolayers, which contain an apical surface and a basolateral surface, it has been shown that the distribution of six glycosyl-PI-anchored proteins is restricted to the apical surface, suggesting a role for glycosyl-PI in targeting proteins to specific sites of the membrane.[108] The contact site A glycoprotein of *Dictyostelium discoideum*, an adhesion protein expressed at the aggregation stage of this organism, has been labeled with *myo*-inositol, fatty acids, phosphate, and ethanolamine *in vivo*.[110] However, this lipid was not sensitive to PI-PlC treatment, but, rather, to digestion with sphingomyelinase, suggesting the existence of a novel type of phospholipid anchor.[110] Moreover, these results indicate that lack of sensitivity to PI-PlC cannot be used to exclude the possibility that a membrane protein contains a phospholipid anchor. In hepatocytes, heparan sulfate proteoglycans have been reported to contain a glycosyl-PI anchor.[68] However, results from other investigators suggest that heparan sulfate proteoglycans are not attached to the membrane via a glycosyl-PI anchor.[115]

The proteins listed in Table 1 belong to three main categories: (1) hydrolytic enzymes, (2) protein coats and adhesion proteins from protozoa, and (3) protein coats, membrane antigens, and adhesion proteins from mammaliam cells. In the first group, there are several proteins, like acetylcholinesterase and alkaline phosphatase, which are abundant in the plasma membrane of many cells and tissues.[19-21] The second group includes several important proteins of simple eukaryotes, such as the variant surface glycoprotein (VSG) of *Trypanosoma brucei*, which is used by the trypanosomes as a protective coat to survive their

TABLE 1
Proteins Which Are Attached to the Membrane Exclusively by Glycosyl-Phosphatidylinositol Anchors

	Ref.
Hydrolytic enzymes	
Acetylcholinesterase	19,21
Alkaline phosphatase	20,21
5′-Nucleotidase	82
Renal dipeptidase	83,124
Trehalase	84
Lipoprotein lipase	71
Proteases	
p63 *Leishmania major*	85,86
p76 *Plasmodium falciparum*	67
Protozoals coats and adhesion proteins	
Paramecium primaurelia surface antigen	87
Trypanosome variant surface glycoprotein	22—25
Dictyostelium 117-kDa protein	26
Mammaliam antigens and adhesion proteins	
Thy-1 antigen	27,28
RT-6 antigen	88
Qa-2 antigen	29,89
Ly-6 antigens	30,31
ThB antigen	29
TAP antigen	81
13- and 52-kDa thymus antigens	90
FcRIII receptor	32—34
Neural cell adhesion molecule, N-CAM	36,122
Heparan sulfate proteoglycan	68
LFA-3 (35)	35
Human carcinoembryonic antigen, CEA	103
P30 antigen of toxoplasma-Gondii	109
Neural cell recognition F11	123
Blast-1 glycoprotein	126
Other proteins	
Decay accelerating factor, DAF	38
Scrapie prion protein	91
Sgp1, Sgp2	92
130-kDa hepatoma glycoprotein	93
158-kDa pheochromocitoma protein	66
34-kDa placental growth factor	37
Folate binding protein	121
Toxoplasma-Gondii major surface antigens	125
Proteins that contain inositol	
Ornitine decarboxylase	17
Myelin basic protein	18

hosts' immune response,[22-25] and the cell-cell cohesion 117-kDa protein of *Dictyostelium discoideum*, which is used by the single amoeba during cell aggregation to form a multi-cellular organism.[26] The third group includes antigens which can trigger cell division upon cross-linking with antibodies, such as Thy-1 and the Ly-6 set of antigens,[27-31] the F_c receptor of phagocytic cells which mediates the binding and clearance of blood immune complexes,[32-34] and cell adhesion proteins such as LFA-3 and the neural protein N-CAM.[35,36] Finally, there is a group of unrelated proteins which are also anchored through a glycosyl-PI. This group includes a 34-kDA growth factor of human placenta[37] and the decay accelerating factor (DAF) of human blood and other tissues which inactivates complement.[38]

It is important to note that all proteins listed in Table 1 do not share any common functional feature besides their type of attachment to the membrane. This suggests that a glycosyl-PI anchor is not related to a specific protein function, and that the information required to direct the attachment of a protein to a glycosyl-PI membrane anchor must be contained within the carboxyl-terminal amino acids of the protein. To test this hypothesis, gene manipulation has been used to construct a fusion protein in which the last 37 amino acids of membrane DAF, which is anchored to the cell surface by a glycosyl-PI anchor,[38] were fused to the carboxyl terminus of a truncated form of glycoprotein D from herpes simplex virus type 1.[39] This truncated glycoprotein is secreted to the medium, since it lacks the carboxyl-terminal membrane-spanning domain. The resultant fusion protein, containing the carboxyl-terminal sequence of DAF, was found to be attached to the membrane by means of a glycosyl-PI anchor, indicating that the signal for attachment to a glycosyl-PI anchor is present at the carboxyl-terminal domain.[39] Similar findings have been reported with the CD8 T-lymphocyte surface antigen using the 3'-end sequence of DAF mRNA.[40] The chimeric protein in transfectants demonstrated glycolipid anchoring, whereas unaltered CD8 in control experiments did not. However, comparison of the carboxyl-terminal amino acid sequences available for PI-anchored proteins has revealed no obvious homology.[10,11,13] The absence of sequence conservation in the carboxyl-terminal region of glycosyl-PI-anchored proteins has led to the suggestion that the information necessary for directing protein attachment to the membrane through a phospholipid anchor might be conformational in character.[39] This hypothesis has been recently tested by showing that replacement of the DAF carboxyl-terminal domain with a secretion signal peptide, or with a random hydrophobic sequence, results in efficient and correct processing of the protein, producing a glycosyl-PI-anchored DAF on the cell surface.[104] During processing of nascent placental alkaline phosphatase, the last 29 amino acids at the carboxyl terminal are removed and the resulting protein is attached to a glycosyl-PI.[100] Using a series of carboxyl-terminal mutants of cDNA encoding human preproplacental alkaline phosphatase, it has been found that in order to be PI anchored, a minimal length of predominantly hydrophobic amino acids is required. While forms of the enzyme with 17 consecutive hydrophobic residues in the terminal sequence were PI anchored and appeared at the cell surface, mutants with 13 or fewer such amino acids were no longer PI anchored.[100] Nascent placental alkaline phosphatase has been expressed in a cell free system to serve as substrate for *in vitro* glycosyl-PI anchoring by microsomal extracts.[111] Microsomal extracts from CHO cells remove the carboxyl-terminal signal peptide from the prepro enzyme in an apparently specific manner, and it has been proposed that this specific cleavage might be the result of the action of a specific transaminidase.[111]

III. STRUCTURE OF THE GLYCOSYL-PI ANCHOR

The general structure of inositol glycophospholipids is given in Figure 1. This is based largely on data for the structure of the glycosyl-PI anchor for the VSG in *T. brucei*, the Thy-1 anchor, and, more recently, the glycosyl-PI antigens of *Leishmania major*. Four different moieties can be distinguished in these glycosyl-PI molecules: (1) the diacylglycerol moiety, (2) the glucosamine-inositol-phosphate group, (3) the glycan moiety, and (4) the phosphoethanolamine bridge.

A. THE DIACYLGLYCEROL MOIETY

The fatty acids of the diacylglycerol group of the various glycosyl-PI anchors show a wide variation. As a rule, many glycosyl-PI anchors contain only or a majority of saturated fatty acids. In the VSGs of *T. brucei* and other African trypanosomes, only myristic acid (C14:0) has been found.[22-24] In *T. equiperdum*, the PI anchor of the VSG has been labeled with palmitic, stearic, and oleic acid.[41] Based on its sensitivity to phospholipase A_2 treatment, the structure of the glycerol backbone in *T. brucei* SVG has been shown to be *sn*-1,2-diacyl-

3-phosphate.[22] The glycosyl-PI of *T. brucei* VSG, however, is resistant to the major phospholipase A$_1$ activity of *T. brucei*, due probably to steric hindrance by the VSG protein or to the structure of the glycan moiety.[42] Myristic and palmitic acid have been found in the 63-kDa protease of *L. major* and in the 195-kDa merozoite antigen of *Plasmodium falciparum*.[13,43] The predominant fatty acid in the glycosyl-PI of higher eukaryotes is steraric acid (C 18:0) although a variety of saturated and unsaturated fatty acids have also been reported. In addition, there is also evidence for the presence of a 1-alkyl-2-acyl structure in the glycerol backbone of the glycosyl-PI anchor of bovine acetylcholinesterase[44,45] and of the glycosyl-PI antigens of *L. major*.[112]

B. THE GLUCOSAMINE-INOSITOL-PHOSPHATE GROUP

Non-*N*-acetylated glucosamine (GlcN) and inositol have been found in all proteins which are known to be glycosyl-PI anchored. Nitrous acid deamination of the glycosyl-PI anchor has been most frequently used to demonstrate the presence of GlcN and is a diagnostic reaction for a glycosyl-PI structure.[10-13] In this reaction, GlcN is converted to 2,5-anhydromannose, with cleavage of the glycosidic bond of the amino sugar and release of phosphatidylinositol.[22-24] These studies demonstrate that the phospho-inositol group of the glycosyl-PI molecule is glycosidically bound to GlcN. Studies with the glycosyl-PI anchors of *T. brucei* VSG and rat brain Thy-1 have shown that the GlcN linkage is in the α-configuration and to the inositol 6-position.[46-48]

By gas chromatography/mass spectrometry, *myo*-inositol has been identified in the glycosyl-PI anchor of a variety of proteins.[11,13] In several other cases, including *Torpedo* acetylcholinesterase and alkaline phosphatase from intestinal and human placenta, *myo*-inositol as well as *chiro*-inositol have been identified.[19,21] The biological importance of the presence of both isomers of inositol has not been established. In addition, the available evidence indicates that no other phosphates (besides the phosphodiester bond between the diacylglycerol and inositol groups) or carbohydrates (except the molecule of GlcN) are associated with the residue of inositol.[10-13]

C. THE GLYCAN MOIETY

The complete structure of the glycan moiety of the glycosyl-PI anchor has been determined for two VSGs of *T. brucei* and for rat brain Thy-1 by a combination of two-dimensional ¹H-NMR and mass spectrometry,[46-48] and for three glycosyl-PI antigens of *L. major* by a combination of gas-liquid chromatography-mass spectrometry and fast atom bombardment-mass spectrometry.[112] The structure of the three molecules is presented in Figure 2. The glycosyl-PI anchor of *T. brucei* has been found to be heterogeneous with respect to the monosaccharide sequence. A group of three mannose residues are either in a linear or branched sequence. This mannose region is further substituted with one α1 → 2-linked galactose branch also showing microheterogeneity.[46,47]

The structure of the glycosyl-PI anchor of rat brain Thy-1 has also been elucidated.[48] It has a mannose backbone identical to one of the VSG anchors reported. The proposed structure of Thy-1 is homogeneous except for one α1→2 mannose residue at the end of the oligosaccharide chain. This molecule also contains one β1→4 *N*-acetyl galactosamine residue linked to the first mannose of the oligosaccharide chain. No detailed information on the glycan moiety of other protein anchors is available. The partial structure of the glycosyl-PI anchor of human erythrocyte acetylcholinesterase has also been reported.[105,106] The data indicate the presence of a plasmanylinositol with a palmitoyl group on an inositol hydroxyl group.[105,106] Chemical analysis of human placental alkaline phosphatase revealed the presence of ethanolamine, glucosamine, mannose, inositol, and phosphate in a ratio of 2:1:3:1:2.[99] From the limited information with *T. brucei* VSG and Thy-1 anchor, the existence of side-chain variations around a relatively constant mannose backbone may be predicted.

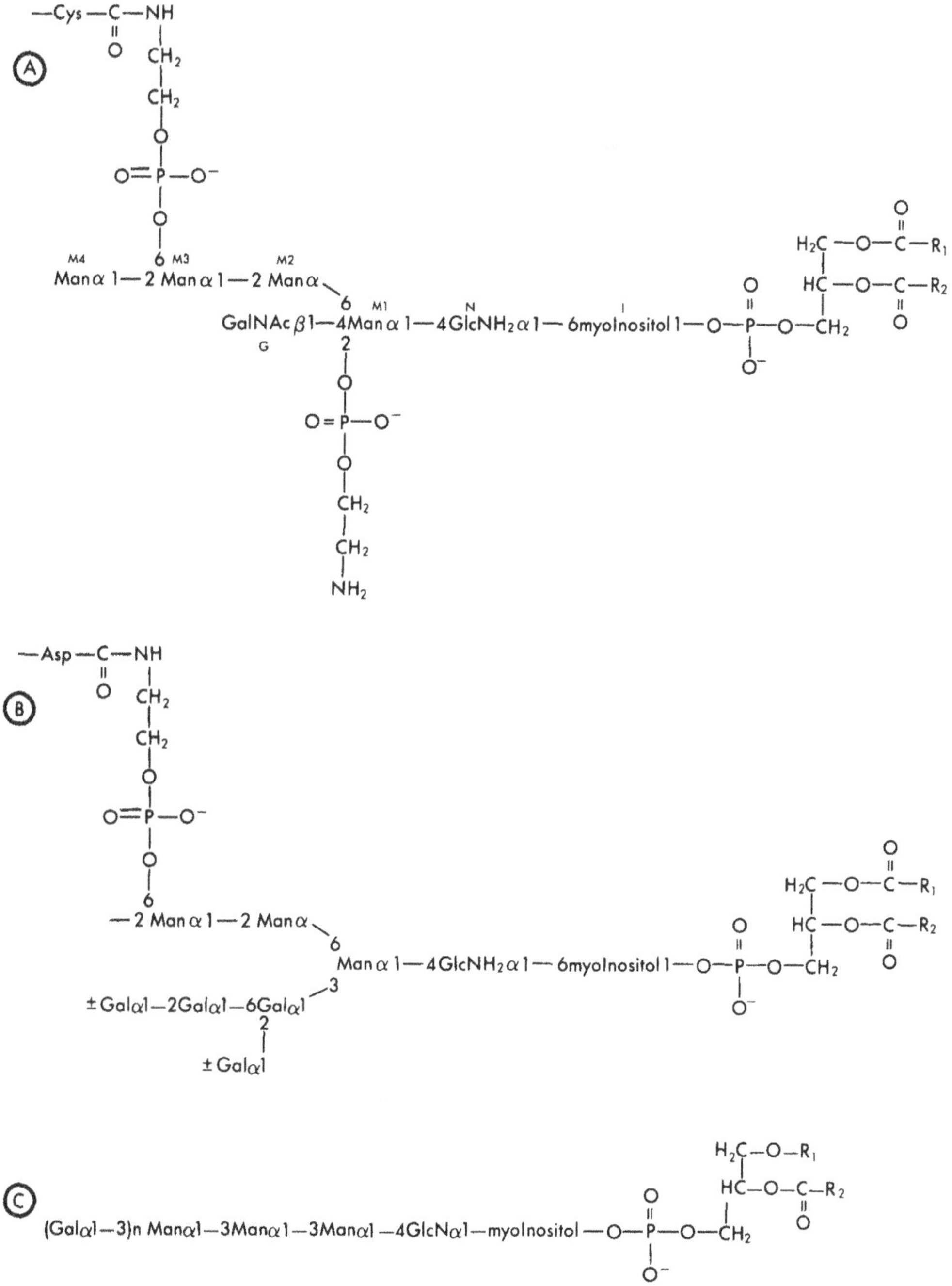

—Cys —C—NH
‖
O CH₂

Ⓐ

CH₂

O

O=P—O⁻

O

M4   M3   M2
Manα 1— 2 Man α 1— 2 Manα
                              6 M1   N
                    GalNAcβ1—4Manα 1—4GlcNH₂α1— 6myoinositol 1—O—P—O—CH₂
                       G   2

$$\begin{array}{l} H_2C—O—C—R_1 \\ HC—O—C—R_2 \end{array}$$

O=P—O⁻

O

CH₂

CH₂

NH₂

—Asp —C—NH
‖
O CH₂

Ⓑ

CH₂

O

O=P—O⁻

O

6
— 2 Man α 1— 2 Manα
                     6
          Manα 1—4GlcNH₂α1— 6myoinositol 1—O—P—O—CH₂
                     3
±Galα1—2Galα1—6Galα1
               2

±Galα1

Ⓒ

(Galα1—3)n Manα1—3Manα1—3Manα1 —4GlcNα1—myoinositol —O—P—O—CH₂

FIGURE 2.   Composite structures of the glycosyl phosphatidylinositol anchors of rat brain Thy-1 (A), *T. brucei* VSG (B), and *L. major* antigens (C). (Structures are redrawn from References 46, 48, and 112.)

D. THE PHOSPHOETHANOLAMINE BRIDGE

In *T. brucei* VSGs, only one ethanolamine residue has been detected per molecule of glycosyl-PI. The hydroxy group of the ethanolamine molecule is linked, via a phosphodiester bond, to the 6-position of the nonreducing terminal mannose residue of the glycan, and the amino group forms an amide bond to the carboxyl-terminal amino acid of the protein.[46,47] In rat brain Thy-1, two ethanolamine residues have been detected per molecule of glycosyl-

PI.[48] As in *T. brucei* VSG, one ethanolamine residue is linked via a phosphodiester bond to the 6-position of the last mannose residue and via an amide bond to the carboxyl-terminal amino acid of the protein. The second residue of ethanolamine in Thy-1 is linked via a phosphodiester bond to the 2-position of the mannose, which is glycosidically bound to the residue of GlcN. This second residue has a free amino group. All glycosyl-PI anchors so far studied, except that from *T. brucei* VSGs, contain two ethanolamine residues, one serving as the protein bridge and the second having the amino groups free.[49-51]

IV. PHOSPHOLIPASES ACTING ON GLYCOSYL-PI ANCHORS

Phospholipases C, which are specific for phosphatidylinositol and cannot hydrolyze a variety of other common phospholipid species (i.e., phosphatidylethanolamine, phosphatidylcholine, phosphatidylinositol-4-phosphate, and phosphatidylinositol-4,5-bisphosphate) have been purified from a variety of bacteria (i.e., *Bacillus cereus*,[52,53] *B. thuringiensis*[53] *Clostridium novyi*,[54] and *Staphylococcus aureus*[55]). The gene for the phospholipase C from *B. thuringiensis* has been isolated and sequenced.[130] No obvious sequence similarities were found between this sequence and that of other phosphoinositide-specific phospholipases C of mammalian origin (see Chapter 7). These enzymes have been shown to catalyze the phosphodiesteratic hydrolysis of many glycosyl-PI anchors and are named PI-PlC (phosphatidylinositol-specific phospholipase C). As a result of the action of these enzymes, glycosyl-PI-anchored membrane proteins are released in a soluble form. This reaction, together with the sensitivity to nitrous acid deamination, are considered to be diagnostic tests of the presence of a glycosyl-PI structure. Several proteins which are PI anchored are, however, partially or completely resistant to PI-PlC. Thus, human acetylcholinesterase, which is PI anchored, is resistant to PI-PlC treatment,[50] and this seems to be due to the presence of one palmitic acid residue in ester linkage to the inositol ring.[105,106] *Drosophila* acetylcholinesterase also contains a glycosyl-PI anchor which is PI-PlC sensitive.[96] The presence of a PI anchor in *Drosophila* is interesting, since all other glycosyl-PI-anchored proteins listed in Table 1 are from mammals or protozoa.

A membrane-bound phospholipase C has been purified from *T. brucei* which seems to be specific for glycosyl-PI anchors and has an M_r of about 40 kDa.[56-58] This enzyme, which shows little activity with phosphatidylinositol, hydrolyzes a variety of glycosyl-PI anchors, including those of *T. brucei* VSG,[24] *Torpedo* acetylcholinesterase,[59] and *Saccharomyces cerevisiae* 125-kDa membrane glycoprotein,[60] to generate the soluble protein and diacylglycerol. A similar protein with an M_r of 52 kDa has been partially purified from rat liver membranes.[61] The glycosyl-PI-specific phospholipase C of *T. brucei* has been cloned and its cDNA sequenced.[101] The encoded polypeptide has an M_r of 40,760 Da, which agrees with previously published data based on electrophoretic mobility.[56-58] Interestingly, there are no similarities between this phospholipase from *T. brucei* and several previously published phosphatidylinositol-specific phospholipases C (see Chapter 7). Furthermore, its deduced amino acid sequence has little similarity with that of other typical membrane proteins, and there are no internal hydrophobic sequences to explain its association with membranes.[101] Cellular localization of the phospholipase C of *T. brucei*, by immunofluorescence microscopy of isolated subcellular organelles and by immunoelectron microscopy using cryosections, indicates that this enzyme is predominantly on the cytoplasmic side of intracellular membranes and is absent from the plasma membrane.[113] These experiments suggest that the phospholipase is separated from its theoretically major substrate, the variant surface glycoprotein. This agrees with the observation that the release of this glycoprotein is slow in cultured *Trypanosoma*.[114] However, these results raise questions about how this enzyme and the variant surface glycoprotein substrate might come together on the same side of the membrane during differentiation or after hipotonic cell lysis.

A phospholipase D cleaving a variety of glycosyl-PI anchors has been identified in mammalian sera[62,63] and purified to homogeneity.[120] This enzyme, with an M_r of 500 kDa, as determined by gel filtration, had no activity on phosphatidylinositol.[62,63] By sodium dodecyl sulfate-polyacrylamide gel electrophoresis, the homogeneous enzyme has an M_r of 110 kDa and seems to consist of a single polypeptide chain.[120]

V. FUNCTIONS OF GLYCOSYL-PI ANCHORS

An obvious function of glycosyl-PI anchors is to attach proteins to the membrane. Proteins attached to the membrane through PI anchors require detergents for extraction, and in this way behave like integral membrane proteins. However, whereas integral proteins cross the lipid bilayer, via a hydrophobic protein sequence, exposing two hydrophylic moieties, one at the outer surface and the other at the cytoplasmic side of the cell, glycosyl-PI-anchored proteins do not cross the lipid bilayer and are only exposed to the outer surface of the cell (Figure 1). At this point, it is interesting to note that certain membrane proteins are modified by the covalent attachment of long-chain fatty acids. Contrary to the situation with PI anchors, the majority of these proteins are localized in the cytoplasmic surface of the plasma membrane, and they raise the possibility that fatty acid acylation may play a role in intracellular sorting of nontransmembranous, nonglycosylated membrane proteins.[98] Proteins with glycosyl-PI anchors should behave more like phospholipid molecules having a polar headgroup consisting of several thousands daltons. This might facilitate the lateral mobility of the anchored protein, although interactions with other membrane proteins, which may restrict the diffusion rate, are also likely to occur. Thus, fractions of Thy-1 show diffusion coefficients considerably larger than those of many other plasma membrane proteins (diffusion coefficients around 2 to 4 $\times$ 10^{-9} cm²/s) and similar to those of lipids, although part of the Thy-1 wa also not mobile.[64] PH-20, a glycosyl-PI-anchored sperm surface protein involved in sperm-egg adhesion, has been shown to have a highly restricted diffusion rate, being over 1000-fold slower than lipid diffusion.[65] Since glycosyl-PI-anchored molecules do not contain a cytoplasmic domain, these proteins do not communicate with the interior of the cell, which limits their function to the extracellular space. In this respect, molecules with PI anchors are similar to proteins excreted by the cell to the medium. However, these molecules with PI-anchors are not free to diffuse in the extracellular medium, but, rather, concentrate around the cell surface. This might be an advantage in the case of hydrolytic enzymes like acetylcholinesterase, where rapid hydrolysis of acetylcholine is desirable for efficient neurotransmission. In Madin-Darby canine kidney cells, which form polarized monolayers at confluency, the vesicular stomatitis virus glycoprotein (VSV G) is a trans-membrane protein normally localized in the basolateral plasma membrane domain, and the alkaline phosphatase is present in the apical surface. The conversion of the normal VSV G to a glycosyl-PI-anchored protein shifted its expression from the basolateral to the apical domain, and the conversion of the placental alkaline phosphatase with the transmembrane and cytoplasmic domains of VSV G switched its expression from the apical to the basolateral surface.[118] These results suggest that the glycosyl-PI anchor is a signal for the apical transport of proteins.

Treatment of intact cells with exogenous PI-PlC results in the hydrolysis of the phos-phodiester bond that links PI-anchored proteins to the cell, with generation of the soluble protein. For instance, incubation of PC12 pheochromocytoma cells with PI-PlC results in the release of a broad band with an M_r of 32 kDa, probably Thy-1 glycoprotein, and a second protein of 158 kDa.[66] The existence of membrane-bound PI-PlCs, and the presence of phospholipase D activity in the sera of a variety of species, suggests that phospholipase hydrolysis might be a physiological process to regulate the concentration of PI-anchored proteins at the cell surface. This might be of particular utility in the case of protective-coat proteins in parasites and for proteins involved in cell adhesion, recognition, etc., several of which have been shown to be PI-anchored (Table 1). Furthermore, this type of cleavage might change the conformation of the protein, by exposing a glycan moiety in the soluble protein, which might result in the generation of a novel enzymatic activity or a specific site for binding to the same or another cell. Evidence in favor of this hypothesis has been obtained by showing that in *Plasmodium falciparum*, PI-PlC releases a membrane protein with an M_r of 76 kDa from intact merozoites or isolated schizont membranes and induces a proteolytic activity associated only with its soluble form.[67] The finding that endogenous activation of this proteolytic activity occurs at the end of the schizogony[67] suggests that the induction of this activity by PI-PlC might be of importance in merozoite maturation. In hepatocytes, evidence has been given indicating that the heparan sulfate proteoglycan which is newly secreted is initially bound to the plasma membrane through a glycosyl-PI anchor. In the presence of insulin, the glycosyl-PI-anchored proteoglycan is cleaved on the cell surface (probably by an insulin-sensitive phospholipase C), then to be internalized via binding to a glycosyl-inositol phosphate "receptor".[68] Insulin also has been shown to decrease the cellular levels of alkaline phosphatase[69,70] and lipoprotein lipase,[71] two proteins which are thought to be PI anchored, probably by activating a membrane-bound insulin-sensitive phospholipase C having its active site at the cell surface.

Three types of receptor for the constant (Fc) region of human immunoglobulin have been described: FcRI, FcRII, and FcRIII. FcRIII (CD16) is a low-affinity receptor expressed in macrophages, neutrophils, eosinophils, natural killer cells, and a subset of T cells believed to comprise the suppressor cells. FcRIII, which accounts for most of the FcR in blood, has been shown to have a PI anchor in neutrophils, whereas in natural killer cells and macrophages, it is found as a transmembrane-anchored molecule.[32-34] A serine residue, Ser 203, in the glycosyl-PI attachment domain seems to determine whether the molecule of FcRIII is PI anchored.[127-129] Two similar genes, CD16-I and CD16-II, encode membrane glycoproteins that are anchored by glycosyl-PI and transmembrane polypeptides, respectively. Although both cDNAs have an identical carboxy-terminal domain, CD16-I has a serine residue in position 203, whereas CD16-II has phenylalanine in this same position. Conversion of Phe to Ser in CD16-II permits expression of a glycosyl-PI-anchored FcRIII, whereas conversion of Ser to Phe in CD16-I prevents glycosyl-PI anchoring.[129] The expression of FcRIII has been shown to be deficient in paroxysmal nocturnal hemoglobinuria (PNH),[32,34] an acquired abnormality of hematopoietic cells affecting the expression of PI-anchored proteins.[36] This deficiency of FcRIII in PNH seems to be responsible for circulating immuno complexes and the susceptibility to bacterial infections associated with this disease. FcRIII also has been implicated in receptor-mediated endocytosis and in triggering cell-mediated killing. The synthesis of FcRIII in PNH patients appears to be normal,[34] indicating that the defect lies in the glycosyl-PI tail. The question is how this glycosyl-PI-anchored protein can mediate ligand internalization and cytotoxicity. FcRIII release has been observed upon stimulation of neutrophils by the inflammatory bacterial peptide f-Met-Leu-Phe, suggesting a role for FcRIII release in inflammatory reactions.[34] Whether the released FcRIII binds to the cell surface, as mentioned above for the heparan sulfate proteoglycan in hepatocytes, and this is important to its role in ligand internalization and cytotoxicity, remains to be determined. FcRIII from human natural killer cells is insensitive to PI-PlC,[125] indicating the existence of different isoforms with different types of membrane anchors.

Antibodies specific against Qa-2, a glycosyl-PI-anchored class I histocompatibility antigen, activate mouse T cells. To determine whether the glycosyl-PI anchor is important for cell activation, transgenic mice expressing either normal glycosyl-PI-anchored Qa-2 or Qa-2 molecules with a transmembrane-spanning domain derived from H-2 have been produced.[119] Interestingly, only lymphocytes from transgenic mice carrying glycosyl-PI-anchored forms of Qa-2 can be activated *in vitro* by Qa-2-specific antibodies. Moreover, lymphocytes from transgenic animals carrying a glycosyl-PI-anchored form of H-2D were activated by anti-H-2D antibodies. These results support the view that glycosyl-PI anchors can play an essential role during cell activation.

Phospholipase C hydrolysis of the phosphodiester bond of a glycosyl-PI not only results in the release of the protein, but also in the generation of diacylglycerol at the outer layer of the plasma membrane, or of phosphatidic acid if the hydrolytic enzyme has phospholipase D activity. Other possibilities include (1) the hydrolysis of the PI anchor by an endoglycosidase, which results in the generation of a glycophospholipid, and (2) the combined action of a protease and phospholipase C, which results in the production of a phospho-oligosaccharide, as has been suggested in the case of alkaline phosphatase.[70] It is interesting to note that all these molecules (diacylglycerol, phosphatidic acid, glycophospholipids, and phospho-oligosaccharides) have been implicated as cellular signaling mediators,[72-74] which opens a new field of functions for glycosyl-PI lipids (see Chapter 8).

VI. BIOSYNTHESIS OF GLYCOSYL-PI ANCHORS

Attachment of the VSG or Thy-1 C-terminals to the membrane is very rapid, less than 2 min upon translation of the protein.[75-77] This posttranslational processing consists of two steps. In the first step, a hydrophobic peptide sequence, consisting of about 20 to 30 amino acids, is removed from the C terminus; and in the second step, the resulting protein is coupled to the glycosyl-PI anchor. In QGP-1 cells derived from a human pancreatic carcinoma, the human carcinoembryonic antigen is PI anchored to the membrane immediately after synthesis by simultaneously occurring proteolysis of the carboxyl terminal and substitution by the glycosyl-PI.[103] In *T. brucei* VSG, the glycosyl-PI anchor is added in the endoplasmic reticulum, to be subsequently transported along the classical intracellular route for glycoproteins.[77-79] The speed of the process suggests the existence of a pool of glycosyl-PI anchors ready to be coupled to the protein as soon as the C-terminal peptide is removed. In *T. brucei*, a glycophospholipid has been isolated which might be the preassembled glycosyl-PI.[5] Inositol glycophospholipids with the general structure lipid-phosphate-inositol-glycan have been isolated from plants,[2] bacteria,[3] yeast,[4] protozoa,[5,6] and mammaliam cells.[7-9] Whether any of these glycosphospholipids serve as preassembled glycosyl-PI anchors remains to be determined. The enzymes involved in the biosynthesis of glycosyl-PI anchors are unknown, as are the enzymes for attachment of the protein to the ethanolamine group. Glycosyl-PI anchors seem to be synthesized by sequential glycosylation of phosphatidylinositol. The first step in the pathway is the transfer of *N*-acetylglucosamine from UDP-*N*-acetylglucosamine to endogenous phosphatidylinositol to form *N*-acetylglucosaminyl-PI.[116,117] This lipid is then deacylated to form glucosaminyl-PI, which is further used for glyconjugation reactions.[117] In *L. major*, three glycosyl-PIs have been identified whose structures agree with being sequential steps in the biosynthesis of glycosyl-PI anchors.[112] A variety of cell mutants have been isolated in which the Thy-1 protein or TAP, a 10 to 12-kDa glycoprotein that is expressed on the plasma membrane of mature T lymphocytes, are synthesized, but the protein does not appear at the cell surface, suggesting defects in the cleavage of the protein or the attachment of the glycosyl-PI group.[51,77,80,81] In PNH, there is also a deficiency in the synthesis or transfer of the glycosyl-PI anchor to proteins.[32,34,36] Similarly, in mutants that lack expression of glycosyl-PI-anchored proteins, only the transmembrane form of LFA-3 was detected.[107] These results suggest that the biosynthesis of glycosyl-PI anchors might be a complex process.

REFERENCES

1. **Houslay, M. D. and Stanley, K. K.,** *Dynamics of Biological Membranes,* John Wiley & Sons, New York, 1982, chap. 3.
2. **Carter, H. E., Strobach, D. R., and Hawthorne, J. N.,** Biochemistry of the sphingolipids. XVIII. Complete structure of tetrasaccharide phytoglycolipid, *Biochemistry,* 8, 383, 1969.
3. **Lee, Y. C. and Ballou, C. E.,** Complete structures of the glycophospholipids of mycobacteria, *Biochemistry,* 4, 1395, 1965.
4. **Steiner, S., Smith, S., Waechter, C. J., and Lester, R. L.,** Isolation and partial characterization of a major inositol-containing lipid in baker's yeast mannosyl-diinositol, diphosphoryl-ceramide, *Proc. Natl. Acad. Sci. U.S.A.,* 64, 1042, 1969.
5. **Krarkow, J. L., Hereld, D., Bangs, J. D., Hart, G. W., and Englund, P. T.,** Identification of a glycolipid precursor of the *Trypanosoma brucei* variant surface glycoprotein, *J. Biol. Chem.,* 261, 12147, 1986.
6. **Menon, A. K., Mayor, S., Ferguson, M. A. J., Duszenko, M., and Cross, G. A. M.,** Candidate glycophospholipid precursor for the glycosylphosphatidylinositol membrane anchor of *Trypanosoma brucei* surface glycoprotein, *J. Biol. Chem.,* 263, 1970, 1988.
7. **Saltiel, A. R. and Cuatrecasas, P.,** Insulin stimulates the generation from hepatic plasma membranes of modulators derived from an inositol glycolipid, *Proc. Natl. Acad. Sci. U.S.A.,* 83, 5793, 1986.
8. **Saltiel, A. R., Fox, J. A., Sherline, P., and Cuatrecasas, P.,** Insulin stimulated hydrolysis of a novel glycolipid generates modulators of cAMP phosphodiesterase, *Science,* 233, 967, 1986.
9. **Mato, J. M., Kelly, K. L., Abler, A., and Jarett, L.,** Identification of a novel insulin-sensitive glyco-phospholipid from H35 hepatoma cells, *J. Biol. Chem.,* 262, 2131, 1987.
10. **Cross, G. A. M.,** Eukaryotic protein modification and membrane attachment via phosphatidylinositol, *Cell,* 48, 179, 1987.
11. **Low, M. G.,** Biochemistry of the glycosyl-phosphatidylinositol membrane protein anchors, *Biochem. J.,* 244, 1, 1987.
12. **Low, M. G. and Saltiel, A. R.,** Structural and functional roles of glycosyl-phosphatidylinositol in membranes, *Science,* 239, 268, 1988.
13. **Ferguson, M. A. J. and Williams, A. F.,** Cell-surface anchoring of proteins via glycosyl-phosphatidy-linositol structures, *Annu. Rev. Biochem.,* 57, 285, 1988.
14. **Mato, J. M.,** Insulin mediators revisited, *Cell. Signal.,* 1, 143, 1989.
15. **Slein, M. W. and Logan, G. F.,** Mechanism of action of the toxin of *Bacillus anthracis.* I. Effect in vivo on some blood serum components, *J. Bacteriol.,* 80, 77, 1960.
16. **Slein, M. W. and Logan, G. F.,** Characterization of the phospholipases of *Bacillus cereus* and their effects on erythrocyte, bone and kidney cells, *J. Bacteriol.,* 90, 69, 1965.
17. **Mustelin, T., Pösö, H., Lapinjoki, S. P., Gynther, J., and Andersson, L. C.,** Growth signal transduction: rapid activation of covalently bound ornitine decarboxylase during phosphatidylinositol breakdown, *Cell,* 49, 171, 1987.
18. **Yang, Y. C., Chang, P. C., Fujitaki, J. M., Chiu, K. C., and Smith, R. A.,** Covalent linkage of phospholipid to myelin basic protein: identification of phosphatidylinositol bisphosphate as the attached phospholipid, *Biochemistry,* 25, 2677, 1986.
19. **Futerman, A. H., Low, M. G., Ackerman, K. E., Sherman, W. R., and Silman, I.,** Biochemical behaviour and structural characteristics of membrane-bound acetylcholinesterase from *Torpedo* electric organ, *Biochem. Biophys. Res. Commun.,* 129, 312, 1985.
20. **Kominami, T., Miki, A., and Ikehara, Y.,** Electrophoretic characterization of hepatic alkaline phosphatase released by phosphatidylinositol-specific phospholipase C, *Biochem. J.,* 227, 183, 1985.
21. **Low, M. G., Futerman, A. H., Ackerman, K. E., Sherman, W. R., and Silman, I.,** Removal of covalently bound inositol from *Torpedo* acetylcholinesterase and mammalian alkaline phosphatase by deam-ination with nitrous acid, *Biochem. J.,* 241, 615, 1987.
22. **Ferguson, M. A. J. and Cross, G. A. M.,** Myristylation of the membrane form of a *Trypanosome brucei* variant surface glycoprotein, *J. Biol. Chem.,* 259, 3011, 1984.
23. **Ferguson, M. A. J., Haldar, K., and Cross, G. A. M.,** *Trypanosoma brucei* variant surface glycoprotein has an sn-1,2-dimyristyl glycerol membrane anchor at its COOH terminus, *J. Biol. Chem.,* 260, 4963, 1985.
24. **Ferguson, M. A. J., Low, M. G., and Cross, G. A. M.,** Glycosyl-sn-1,2-dimyristylphosphatidylinositol is covalently linked to *Trypanosoma brucei* variant surface glycoprotein, *J. Biol. Chem.,* 260, 14547, 1985.
25. **Strang, A. M., Williams, J. M., Ferguson, M. A. J., Holder, A. A., and Allen, A. K.,** *Trypanosoma brucei brucei* variant surface glycoprotein contains non-N-acetylated glucosamine, *Biochem. J.,* 234, 481, 1986.
26. **Sadeghi, H., Silia, A. M., and Klein, C.,** Evidence that a glycolipid tail anchors antigen 117 to the plasma membrane of *Dictyostelium discoideum, Proc. Natl. Acad. Sci. U.S.A.,* 85, 5512, 1988.

27. **Low, M. G. and Kincade, P. W.,** Phosphatidylinositol is the membrane anchoring domain of the Thy-1 glycoprotein, *Nature,* 318, 62, 1985.
28. **Tse, A. G. D., Barclay, A. N., Watts, A., and Williams, A. F.,** A glycophospholipid tail at the carboxyl terminus of the Thy-1 glycoprotein of neurons and thymocytes, *Science,* 230, 1003, 1985.
29. **Stiernberg, J., Low, M. G., Flaherty, L., and Kincade, P. W.,** Removal of lymphocyte surface molecules with phosphatidylinositol specific phospholipase C: effects of mitogen response and evidence that ThB and certain Qa antigens are membrane-anchored via phosphatidylinositol, *J. Immunol.,* 38, 3877, 1987.
30. **Reiser, H., Oetgen, H., Yeh, E. T. H., Terhorst, C., Low, M. G., Benacerraf, B., and Rock, K. L.,** Structural characterization of the TAP molecule: a phosphatidylinositol-linked glycoprotein distinct from the T cell receptor/T3 complex and Thy-1, *Cell,* 47, 365, 1986.
31. **Hammelburger, J. W., Palfree, R. G. E., Sirlin, S., and Hämmerling, U.,** Demonstration of phosphatidylinositol anchors on Ly-6 molecules by specific phospholipase C digestion and gel electrophoresis, *Biochem. Biophys. Res. Commun.,* 148, 1304, 1987.
32. **Selvaraj, P., Rosse, W. F., Silber, R., and Springer, T. A.,** The major Fc receptor in blood has a phosphatidylinositol anchor and is deficient in paroxysmal nocturnal haemoglobinuria, *Nature,* 333, 565, 1988.
33. **Simmons, D. and Seed, B.,** The Fc receptor of natural killer cells is a phospholipid-linked membrane protein, *Nature,* 333, 568, 1988.
34. **Huizinga, T. W., van der Schoot, C. E., Jost, C., Klaassen, R., Kleijer, M., von dem Borne, A. E. G. K., Roos, D., and Tetteroo, P. A. T.,** The PI-linked receptor FcRIII is released on stimulation of neutrophils, *Nature,* 333, 667, 1988.
35. **He, H. T., Barbet, J., Chaix, J. C., and Goridis, C.,** Phosphatidylinositol is involved in the membrane attachment of NCAM-120, the smallest component of the neural cell adhesion molecule, *EMBO J.,* 5, 2489, 1986.
36. **Selvaraj, P., Dustin, M. J., Silber, R., Low, M. G., and Springer, T. A.,** Deficiency of lymphocyte function-associated antigen 3 (LFA-3) in paroxysmal nocturnal hemoglobinuria, *J. Exp. Med.,* 166, 1011, 1987.
37. **Choudhury, S. R., Mishra, V. S., Low, M. G., and Das, M.,** A phospholipid is the membrane-anchoring domain of a protein growth factor of molecular mass 34 kDa in placental trophoblasts, *Proc. Natl. Acad. Sci. U.S.A.,* 85, 2014, 1988.
38. **Davitz, M. A., Low, M., and Nussenzweig, V.,** Release of decay-accelerating factor from the cell membrane by phosphatidylinositol specific phospholipase C, *J. Exp. Med.,* 161, 1150, 1986.
39. **Caras, I., Weddell, G. N., Davitz, M. A., Nussenzweig, V., and Martin, D. W., Jr.,** Signal for attachment of a phospholipid membrane anchor in decay accelerating factor, *Science,* 238, 1280, 1987.
40. **Tykocinski, M. L., Shu, H. K., Ayers, D. J., Walter, E. I., Getty, R. R., Groger, R. K., Hauer, C. A., and Medof, E.,** Glycolipid reanchoring of T-lymphocyte surface antigen CD8 using the 3′ end sequence of decay-accelerating factor's mRNA, *Proc. Natl. Acad. Sci. U.S.A.,* 85, 3555, 1988.
41. **Duvillier, G., Nouvelot, A., Richet, C., Baltz, T., and Degand, P.,** Presence of glycerol and fatty acids in the C-terminal end of a variant surface glycoprotein from *Trypanosoma equiperdum, Biochem. Biophys. Res. Commun.,* 114, 119, 1983.
42. **Hambrey, P. N., Forsberg, C. M., and Mellors, A.,** The phospholipase A_1 of *Trypanosoma brucei* does not release myristate from the variant surface glycoprotein, *J. Biol. Chem.,* 261, 3229, 1986.
43. **Haldar, K., Ferguson, M. A. J., and Cross, G. A. M.,** Acylation of a *Plasmodium falciparum* merozoite surface antigen via *sn-1,2*-diacyl glycerol, *J. Biol. Chem.,* 260, 4969, 1985.
44. **Roberts, W. L., Kim, B. H., and Rosenberry, T. L.,** Differences in the glycolipid membrane anchors of bovine and human erythrocyte acetylcholinesterases, *Proc. Natl. Acad. Sci. U.S.A.,* 84, 7817, 1987.
45. **Roberts, W. L., Myher, J. J., Kuksis, A., and Rosenberry, T. L.,** Alkylacylglycerol molecular species in the glycosylinositol phospholipid membrane anchor of bovine erythrocyte acetylcholinesterase, *Biochem. Biophys. Res. Commun.,* 150, 271, 1988.
46. **Schmitz, B., Klein, R. A., and Duncan, I. A.,** MS and NMR analysis of the cross-reacting determinant glycan from *Trypanosoma brucei* MITat 1.6, *Biochem. Biophys. Res. Commun.,* 146, 1055, 1987.
47. **Ferguson, M. A. J., Homans, S. W., Dwek, R. A., and Rademacher, T. W.,** Glycosyl-phosphatidylinositol moiety that anchors *Trypanosoma brucei* varient surface glycoprotein to the membrane, *Science,* 239, 753, 1988.
48. **Homans, S. W., Ferguson, M. A. J., Dwek, R. A., Rademacher, T. W., Anand, R., and Williams, A. F.,** Complete structure of the glycosyl-phosphatidylinositol membrane anchor of rat brain Thy-1 glycoprotein, *Nature,* 333, 269, 1988.
49. **Haas, R., Brandt, P. T., Knight, J., and Rosenberry, T. L.,** Identification of amine components in a glycolipid membrane-binding domain at the C-terminus of human erythrocyte acetylcholinesterase, *Biochemistry,* 25, 3098, 1986.
50. **Roberts, W. L., Kim, B. H., and Rosenberry, T. L.,** Differences in the glycophospholipid membrane anchors of bovine and human erythrocyte acetylcholinesterases, *Proc. Natl. Acad. Sci. U.S.A.,* 84, 7817, 1987.

51. **Fatemi, S. H., Haas, R., Jentoft, N., Rosenberry, T. L., and Tartakoff, A. M.,** The glycophospholipid anchor of Thy-1. Biosynthetic labeling experiments with wild type and class E Thy-1 negative lymphomas, *J. Biol. Chem.,* 262, 4728, 1987.

52. **Ikezawa, H. and Taguchi, R.,** Phosphatidylinositol-specific phospholipase C from *Bacillus cereus* and *Bacillus thuringiensis, Methods Enzymol.,* 71, 731, 1981.

53. **Taguchi, R. and Ikezawa, R.,** Phosphatidylinositol-specific phospholipase C from *Clostridium novyi* type A, *Arch. Biochem. Biophys.,* 186, 196, 1978.

54. **Ohyabu, T., Taguchi, R., and Ikezawa, H.,** Studies on phosphatidylinositol phosphodiesterase (phospholipase C type) of Bacilus cereum. II. In vivo and immunochemical studies of phosphatase-releasing activity, *Arch. Biochem. Biophys.,* 190, 1, 1978.

55. **Low, M. G.,** Phosphatidylinositol-specific phospholipase C from *Staphylococcus aureus, Methods Enzymol.,* 71, 741, 1981.

56. **Hereld, D., Krakow, J. L., Bangs, J. D., Hart, G. W., and Englund, P. T.,** A phospholipase C from *Trypanosoma brucei* which selectively cleaves the glycolipid on the variant surface glycoprotein, *J. Biol. Chem.,* 261, 13813, 1986.

57. **Fox, J. A., Duszenko, M., Ferguson, M. A. J., Low, M. G., and Cross, G. A. M.,** Purification and characterization of a novel glycan-phosphatidylinositol-specific phospholipase C from *Trypanosoma brucei, J. Biol. Chem.,* 261, 15767, 1986.

58. **Bulow, R. and Overath, P.,** Purification and characterization of the membrane-form variant surface glycoprotein hydrolase of *Trypanosoma brucei, J. Biol. Chem.,* 261, 11918, 1986.

59. **Stieger, A., Cardoso de Almeida, M. L., Blatter, M. C., Brodbeck, U., and Bordier, C.,** The membrane-anchoring systems of vertebrate acetylcholinesterase and variant surface glycoproteins of African trypanosomes share a common antigenic determinant, *FEBS Lett.,* 199, 182, 1986.

60. **Conzelmann, A., Riezman, H., Desponds, C., and Bron, C.,** A major 125-kd membrane glycoprotein of *Saccharomyces cerevisiae* is attached to the lipid bilayer through an inositol-containing phospholipid, *EMBO J.,* 7, 2233, 1988.

61. **Fox, J. A., Soliz, N. M., and Saltiel, A. R.,** Purification of a phosphatidylinositol-glycan-specific phospholipase C from liver plasma membranes: a possible target of insulin action, *Proc. Natl. Acad. Sci. U.S.A.,* 84, 2663, 1987.

62. **Davitz, M. A., Hereld, D., Shak, S., Krakow, J. L., Englund, P. T., and Nussenzweig, V.,** A glycan-phosphatidylinositol-specific phospholipase D in human serum, *Science,* 238, 81, 1987.

63. **Low, M. G. and Prasad, A. R. S.,** A phospholipase D specific for the phosphatidylinositol anchor of cell-surface proteins is abundant in plasma, *Proc. Natl. Acad. Sci. U.S.A.,* 85, 980, 1988.

64. **Ishihara, A., Hou, Y., and Jacobson, K.,** The Thy-1 antigen exhibits rapid lateral diffusion in the plasma membrane of rodent lymphoid cells and fibroblasts, *Proc. Natl. Acad. Sci. U.S.A.,* 84, 1290, 1987.

65. **Phelps, B. M., Primakoff, P., Koppel, D. E., Low, M. G., and Myles, D. G.,** Restricted lateral diffusion of PH-20, a PI-anchored sperm membrane protein, *Science,* 240, 1780, 1988.

66. **Margolis, R. K., Goosen, B., and Margolis, R. U.,** Phosphatidylinositol-anchored glycoproteins of PC12 pheochromocytoma cells and brain, *Biochemistry,* 27, 3454, 1988.

67. **Braun-Breton, C., Rosenberry, T. L., and Pereira da Silva, L.,** Induction of the proteolytic activity of a membrane protein in *Plasmodium falciparum* by phosphatidylinositol-specific phospholipase C, *Nature,* 332, 457, 1988.

68. **Ishihara, M., Fedarko, N. S., and Conrad, E.,** Involvement of phosphatidylinositol and insulin in the coordinate regulation of proteoheparan sulfate metabolism and hepatocyte growth, *J. Biol. Chem.,* 262, 4708, 1987.

69. **Levy, J. R., Murray, E., Manolagas, S., and Olefsky, J. M.,** Demonstration of insulin receptors and modulation of alkaline phosphatase activity by insulin in rat osteoblastic cells, *Endocrinology,* 119, 1786, 1986.

70. **Romero, G., Luttrell, L., Rogol, A., Zeller, K., Hewlett, E., and Larner, J.,** Phosphatidylinositol-glycan anchors of membrane proteins: potential precursors of insulin mediators, *Science,* 240, 509, 1988.

71. **Chan, B. L., Lisanti, M. P., Rodriguez-Boulan, E., and Saltiel, A. R.,** Insulin-stimulated release of lipoprotein lipase by metabolism of its phosphatidylinositol anchor, *Science,* 241, 1670, 1988.

72. **Moolenar, W. H., Kruijer, W., Tilly, B. C., Verlaan, I., Bierman, A. J., and de Laat, S. W.,** Growth factor-like action of phosphatidic acid, *Nature,* 310, 644, 1984.

73. **Nishizuka, Y.,** Turnover of inositol phospholipids and signal transduction, *Science,* 233, 305, 1986.

74. **Alemany, S., Mato, J. M., and Strälfors, P.,** Phospho-dephospho-control by insulin is mimicked by a phospho-oligosaccharide in adipocytes, *Nature,* 330, 77, 1987.

75. **Bangs, J. D., Hereld, D., Krakow, J. L., Hart, G. W., and Englund, P. T.,** Rapid processing of the carboxyl terminus of a trypanosome variant surface glycoprotein, *Proc. Natl. Acad. Sci. U.S.A.,* 82, 3207, 1985.

76. **Bangs, J. D., Andrews, N. W., Hart, G. W., and Englund, P. T.,** Post-translational modification and intracellular transport of a trypanosome variant surface glycoprotein, *J. Cell. Biol.,* 103, 255, 1986.

77. **Conzelmann, A., Spiazzi, A., Hyman, R., and Bron, C.,** Anchoring of membrane proteins via phosphatidylinositol is deficient in two classes of Thy-1-negative mutant lymphoma cells, *EMBO J.*, 246, 605, 1987.

78. **Duszenko, M., Ivanov, I. E., Ferguson, M. A. J., Plesken, H., and Cross, G. A. M.,** Intracellular transport of a variant surface glycoprotein in *Trypanosoma brucei, J. Cell. Biol.*, 106, 77, 1988.

79. **Frevert, U. and Reinwald, E.,** Endocytosis and intracellular occurrence of the variant surface glycoprotein in *Trypanosoma congolense, J. Ultrast. Mol. Struct. Res.*, 99, 137, 1988.

80. **Fatemi, S. H. and Tartakoff, A. M.,** Hydrophilic anchor-deficient Thy-1 is secreted by a class E mutant T lymphoma, *Cell*, 46, 653, 1986.

81. **Yeh, E. T. H., Reiser, H., Bame ai, A., and Rock, K. L.,** TAP transcription and phosphatidylinositol linkage mutants are defective in activation through the T cell receptor, *Cell*, 52, 665, 1988.

82. **Low, M. G. and Finean, J. B.,** Specific release of plasma membrane enzymes by a phosphatidylinositol specific phospholipase C, *Biochim. Biophys. Acta*, 508, 565, 1978.

83. **Hooper, N. M., Low, M. G., and Turner, A. J.,** Renal dipeptidase is one of the membrane proteins released by phosphatidylinositol-specific phospholipase C, *Biochem. J.*, 244, 465, 1987.

84. **Takeshue, Y., Yokota, K., Nishi, Y., Taguchi, R., and Ikezawa, H.,** Solubilization of thralase from rabbit renal and intestinal brush-border membranes by a phosphatidylinositol-specific phospholipase C, *FEBS Lett.*, 201, 5, 1986.

85. **Bordier, C., Etges, R. J., Ward, J., Turner, M. J., and Cardoso de Almeida, M. L.,** Leishmania and trypanosome surface glycoproteins have a common glycophospholipid membrane anchor, *Proc. Natl. Acad. Sci. U.S.A.*, 85, 5988, 1986.

86. **Etges, R., Bouvier, J., and Bordier, C.,** The major surface protein of leishmania promastigotes is anchored in the membrane by a myristic acid-labeled phospholipid, *EMBO J.*, 5, 597, 1986.

87. **Capdeville, Y., Cardoso de Almeida, M. L., and Deregnaucourt, C.,** The membrane anchor of paramecium temperature-specific surface antigens is a glycosylinositol phospholipid, *Biochem. Biophys. Res. Commun.*, 147, 1219, 1987.

88. **Koch, F., Thiele, H. G., and Low, M. G.,** Release of the rat T cell alloantigen RT-6.2 from cell membranes of phosphatidylinositol phospholipase C, *J. Exp. Med.*, 164, 1338, 1986.

89. **Stroynowski, I., Soloski, M., Low, M. G., and Hood, L.,** A single gene encodes soluble and membrane-bound forms of the major histocompatibility Qa-2 antigen: anchoring of the product by a phospholipid tail, *Cell*, 50, 759, 1987.

90. **Waneck, G. L., Sherman, D. H., Kincade, P. W., Low, M. G., and Flavell, R. A.,** Molecular mapping of signals in the Qa-2 antigen required for attachment of the phosphatidylinositol membrane anchor, *Proc. Natl. Acad. Sci. U.S.A.*, 85, 577, 1988.

91. **Pierres, M. and Barbet, J.,** Two novel phospholipid-linked mouse thymocyte surface molecules released by phosphatidylinositol-specific phospholipase C, *Mol. Immunol.*, 24, 1273, 1987.

92. **Stahl, N., Borchelt, D. R., Hsiao, K., and Prusiner, S. B.,** Scrapie prion contains a phosphatidylinositol glycolipid, *Cell*, 51, 229, 1987.

93. **Williams, A. F., Tse, A. G. D., and Gagnon, J.,** Squid glycoproteins with structural similarities to Thy-1 and Ly-6 antigens, *Immunogenetics*, 27, 265, 1988.

94. **Ikehara, Y., Hayashi, Y., Ogata, S., Miki, A., and Kominami, T.,** Purification and characterization of a major glycoprotein in rat hepatoma plasma membranes, *Biochem. J.*, 241, 63, 1987.

96. **Haas, R., Marshall, T. L., and Rosenberry, T. L.,** *Drosophila* acetylcholinesterase: demonstration of a glycoinositol phospholipid anchor and an endogenous proteolytic cleavage, *Biochemistry*, 27, 6453, 1988.

97. **Frederic, J., Malapart, P., Rougon, G., and Barbet, J.,** Cell membrane, but not circulating carcinoembryonic antigen is linked to phosphatidylinositol-containing hydrophobic domain, *Biochem. Biophys. Res. Commun.*, 155, 794, 1988.

98. **Wilcox, C. A. and Olson, E. N.,** The majority of cellular fatty acid acylated proteins are localized to the cytoplasmic surface of the plasma membrane, *Biochemistry*, 26, 1029, 1987.

99. **Ogata, S., Hayashi, Y., Takami, N., and Ikehara, Y.,** Chemical characterization of the membrane-anchoring domain of human placental alkaline phosphatase, *J. Biol. Chem.*, 263, 10489, 1988.

100. **Berger, J., Howard, A. D., Brink, L., Gerber, L., Hauber, J., Cullen, B. R., and Udenfriend, S.,** COOH-terminal requirements for the correct processing of a phosphatidylinositol-glycan anchored membrane protein, *J. Biol. Chem.*, 263, 10016, 1988.

101. **Hereld, D., Hart, G. W., and Englund, P.,** cDNA encoding the glycosyl-phosphatidylinositol-specific phospholipase C of *Trypanosoma brucei, Proc. Natl. Acad. Sci. U.S.A.*, 85, 8914, 1988.

102. **Fouchier, F., Bastiani, P., Baltz, T., Aunis, D., and Rougon, G.,** Glycosylphosphatidylinositol is involved in the membrane attachment of proteins in granules of chromaffin cells, *Biochem. J.*, 256, 103, 1988.

103. **Takami, N., Misumi, Y., Kuroki, M., and Ikehara, Y.,** Evidence for carboxyl-terminal processing and glycolipid-anchoring of human carcinoembryonic antigen, *J. Biol. Chem.*, 263, 12716, 1988.

104. **Caras, I. W. and Weddell, G. N.**, Signal peptide for protein secretion directing glycophospholipid membrane anchor attachment, *Science*, 243, 1196, 1989.
105. **Roberts, W. L., Myher, J. J., Kuksis, A., Low, M. G., and Rosenberry, T. L.**, Lipid analysis of the glycosinositol phospholipid membrane anchor of human erythrocyte acetylcholinesterase, *J. Biol. Chem.*, 263, 18766, 1986.
106. **Roberts, W. L., Santikarn, S., Reinhold, V., and Rosenberry, T. L.**, Structural characterization of the glycoinositol phospholipid membrane anchor of human erythrocyte acetylcholinesterase by fast atom bombardment mass spectrometry, *J. Biol. Chem.*, 263, 18766, 1988.
107. **Hollander, N., Selvaraj, P., and Springer, T. A.**, Biosynthesis and function of LFA-3 in human mutant cells deficient in phosphatidylinositol-anchored proteins, *J. Immunol.*, 141, 4283, 1988.
108. **Lisanti, M. P., Sargiacomo, M., Graeve, L., Saltiel, A. R., and Rodriguez-Boulan, E.**, Polarized apical distribution of glycosyl-phosphatidylinositol-anchored proteins in a renal epithelial cell line, *Proc. Natl. Acad. Sci. U.S.A.*, 85, 9557, 1988.
109. **Nagel, S. D. and Boothroyd, J. C.**, The major surface antigen, P30, of toxoplasma-Gondii is anchored by a glycolipid, *J. Biol. Chem.*, 264, 5569, 1989.
110. **Stadler, J., Keenan, T. W., Bauer, G., and Gerisch, G.**, The contact site A glycoprotein of *Dictyostelium discoideum* carries a phospholipid anchor of a novel type, *EMBO J.*, 8, 371, 1989.
111. **Bailey, C. A., Gerber, L., Howard, A. D., and Udenfriend, S.**, Processing at the carboxyl terminus of nascent placental alkaline phosphatase in a cell-free system: evidence for specific cleavage of a signal peptide, *Proc. Natl. Acad. Sci. U.S.A.*, 86, 22, 1989.
112. **Rosen, G., Pahlsson, P., Londner, M. V., Westerman, M. E., and Nilsson, B.**, Structural analysis of glycosyl-phosphatidylinositol antigens of *Leishmania major*, *J. Biol. Chem.*, 264, 10457, 1989.
113. **Bülow, R., Griffiths, G., Webster, P., Stierhof, Y. D., Opperdoes, F. R., and Overath, P.**, Intracellular localization of the glycosyl-phosphatidylinositol-specific phospholipase C of *Trypanosoma brucei*, *J. Cell. Sci.*, 93, 233, 1989.
114. **Büllow, R., Nonnengässer, C., and Overath, P.**, Release of the variant surface glycoprotein during differentiation of bloodstream to procyclic forms of *Trypanosoma brucei*, *Parasitology*, 32, 85, 1989.
115. **Brandan, E. and Hirschberg, C. B.**, Differential association of rat liver heparan sulfate proteoglycans in membranes of the Golgi apparatus and the plasma membrane, *J. Biol. Chem.*, 264, 10520, 1989.
116. **Masterson, W. J., Doering, T. L., Hart, G. W., and Englund, P. T.**, A novel pathway for glycan assembly: biosynthesis of the glycosyl-phosphatidylinositol anchor of the trypanosome variant surface glycoprotein, *Cell*, 56, 793, 1989.
117. **Doering, T. L., Masterson, W. J., Englund, P. T., and Hart, G. W.**, Biosynthesis of the glycosyl phosphatidylinositol membrane anchor of the trypanosome variant surface glycoprotein, *J. Biol. Chem.*, 264, 11168, 1989.
118. **Brown, D. A., Crise, B., and Rose, J. K.**, Mechanism of membrane anchoring affects polarized expression of two proteins in MDCK cells, *Science*, 245, 1499, 1989.
119. **Robinson, P. J., Millrain, M., Antoniou, J., Simpson, E., and Mellor, A. L.**, A glycophospholipid anchor is required for Qa-2-mediated T cell activation, *Nature*, 342, 85, 1989.
120. **Davitz, M. A., Hom, J., and Schenkman, S.**, Purification of a glycosyl-phosphatidylinositol-specific phospholipase D from human plasma, *J. Biol. Chem.*, 264, 13760, 1989.
121. **Lacey, S. W., Sanders, J. M., Rothberg, K. G., Anderson, R. G., and Kamen, B. A.**, Complementary DNA for the folate binding protein correctly predicts anchoring to the membrane by glycosyl-phosphatidylinositol, *J. Clin. Invest.*, 84, 715, 1989.
122. **Kolbinger, F., Schwarz, K., Brombacher, F., von Kleist, S., and Grunert, F.**, Expression of an NCA cDNA in NIH/3T3 cells yields a 110K glycoprotein, which is anchored into the membrane via glycosyl-phosphatidylinositol, *Biochem. Biophys. Res. Commun.*, 161, 1126, 1989.
123. **Wolff, J. M., Brummendorf, T., and Rathjen, F. G.**, Neural cell recognition molecule F11: membrane interaction by covalently attached phosphatidylinositol, *Biochem. Biophys. Res. Commun.*, 161, 931, 1989.
124. **Littlewood, G. M., Hooper, N. M., and Turner, A. J.**, Ectoenzymes of the kidney microvillar membrane. Affinity purification, characterization and localization of the phospholipase C-solubilized form of renal dipeptidase, *Biochem. J.*, 257, 361, 1989.
125. **Edberg, J. C., Redecha, P. B., Salmon, J. E., and Kimberly, R. P.**, Human Fc gamma RIII (CD16). Isoforms with distinct allelic expression, extracellular domains, and membrane linkages on polymorphonuclear and natural killer cells, *J. Immonol.*, 143, 1642, 1989.
126. **Tomavo, S., Schwarz, R. T., and Dubremetz, J. F.**, Evidence for glycosyl-phosphatidylinositol anchoring of *Toxoplasma-Gondii* major surface antigens, *Mol. Biol. Genet.*, 9, 4576, 1989.
127. **Kurosaki, T. and Ravetch, J. V.**, A single amino acid in the glycosyl phosphatidylinositol attachment domain determines the membrane topology of FcϒIII, *Nature*, 342, 805, 1989.
128. **Hibbs, M. L., Selvaraj, P., Carpen, O., Springer, T. A., Kuster, H., Jouvin, M. H. E., and Kinet, J. P.**, Mechanisms for regulation expression of membrane isoforms of FcϒRIII (CD16), *Science*, 246, 1608, 1989.

129. **Lanier, L. L., Cwirla, S., Yu, G., Testil, R., and Phillips, J. H.,** Membrane anchoring of a human IgG Fc receptor (CD16) determined by a single amino acid, *Science,* 246, 1611, 1989.
130. **Enner, D. J., Yang, M., Chen, E., Hellmiss, R., Rodriguez, H., and Low, M. G.,** Sequence of the *Bacillus thuringiensis* phosphatidylinositol specific phospholipase C, *Nucleic Acid Res.,* 16, 10383, 1988.

Chapter 6

PHOSPHOLIPID METABOLISM AND TURNOVER: PHOSPHATIDIC ACID, PHOSPHATIDYLCHOLINE, AND PHOSPHATIDYLETHANOLAMINE

José M. Mato

TABLE OF CONTENTS

I. GENERAL COMMENTS

Figure 1 summarizes the pathways of biosynthesis of phospholipids and triacylglycerol in mammalian tissues. Essentially, these pathways have been known for over 25 years. The endoplasmic reticulum is the major site for the initial biosynthesis of phospholipids. As mentioned in Chapter 2, in mammals, the various subcellular membranes differ widely in phospholipid composition, and a specific sorting and transport mechanism of phospholipids must therefore exist to guarantee the lipid composition of each organelle within the cell. Phospholipid transfer proteins probably play an important role in the intracellular transfer of lipids between membranes. Phospholipid turnover, which is essential for proper cellular function, is very rapid in most mammalian cells, and about half of the total cellular phospholipid is degraded every one or two cell divisions.[1] From the scheme in Figure 1, it is obvious that phosphatidic acid occupies a key position in the biosynthesis of lipids, since it functions as a precursor for the synthesis of both phospholipids and neutral glycerides. A *second important point that can be made from Figure 1 is that some of the major phospholipids* can be synthesized by modification of the polar headgroup of a preexisting lipid. Thus, phosphatidylethanolamine can be converted into phosphatidylcholine by three successive N-methylations of the amino group, phosphatidylserine can be converted into phosphatidylethanolamine by decarboxylation, and phosphatidylinositol can be converted into phosphatidylinositol-4,5-bisphosphate by phosphorylation of the inositol ring.

A recent advance in the field of lipid research has been the discovery that certain minor phospholipids play an essential role in cell signaling. The first of these started with the observation by Hokin and Hokin in 1953 of an agonist-stimulated phosphatidylinositol turnover.[3] This has led to the description of two types of novel cellular messengers: diacylglycerol and a variety of phosphorylated derivatives of inositol that play a key role in the regulation of normal and transformed cell functions.[3,4] More recently, phosphatidic acid has been shown to stimulate growth,[5] and free long-chain bases, such as sphingosine, have been found to have direct effects on cells as inhibitors of protein kinase C (a diacylglycerol-dependent protein kinase).[6] Gangliosides have also been reported to modulate protein phosphorylation,[7] and the polar headgroup of a novel glycosyl phosphatidylinositol, related to those used by the cell for protein anchoring, has been shown to mimic insulin action on protein phosphorylation/dephosphorylation.[8]

Another surprising development has been the identification of 1-*O*-alkyl-2-*O*-acetyl-*sn*-glycero-3-phosphocholine as a potent mediator of a variety of biological and pathophysiological processes, such as platelet aggregation, hypertension, smooth muscle contraction, and neuronal development.[9]

The finding that phospholipids play a key role in cellular signaling and the identification of novel "minor" phospholipids as biologically active substances led many biochemists to move into the field of phospholipid biochemistry.

II. BIOSYNTHESIS OF PHOSPHATIDIC ACID

As mentioned above, phosphatidic acid plays a key role in the metabolism of glycerolipids, since it is the precursor of both phospholipids and neutral glycerides (Figure 1). Phosphatidic acid is the simplest form of phospholipid, and its polar headgroup consists of only the phosphate group. The biosynthesis of phosphatidic acid occurs via (1) the sequential acylation of glycerol-3-phosphate, (2) the acylation of dihydroxyacetone phosphate, and (3) the diacylglycerol kinase pathway (Figure 2).

Most of the glycerol-3-phosphate needed for lipid synthesis is derived from glucose via the reduction of dihydroxyacetone phosphate, an intermediate of glycolysis. This reaction is catalyzed by glycerol-3-phosphate dehydrogenase, a cytosolic enzyme that uses NADH

63

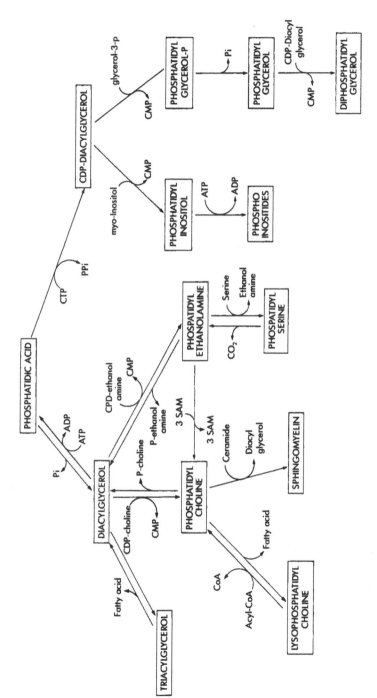

FIGURE 1. Schematic representation of the pathway of biosynthesis of the major phospholipids and triacylglycerol in mammalian tissues.

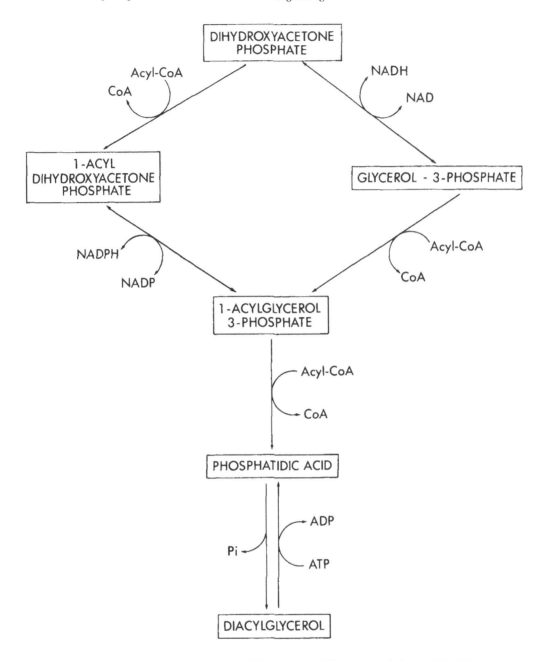

FIGURE 2.   Schematic representation of the pathways of biosynthesis of phosphatidic acid.

as the reductant. Glycerol-3-phosphate can also be synthesized by the action of glycerokinase, an enzyme that catalyzes the phosphorylation of glycerol, using ATP as phosphate donor. However, many tissues cannot use exogenous glycerol and depend on the reduction of dihydroxyacetone phosphate as their only source of glycerol-3-phosphate. Glycerolkinase activity is highest in liver and kidney, yet in these tissues dihydroxyacetone phosphate is probably the main source of glycerol-3-phosphate.[10]

The biosynthesis of phosphatidic acid occurs via the sequential acylation of glycerol-3-phosphate by long-chain fatty acyl-CoA, lysophosphatidic acid being the intermediate of this pathway (Figure 2).[11,12] Lysophosphatidic acid can also be synthesized by acylation of dihydroxyacetone phosphate and the subsequent reduction of the resulting acyldihydroxy-acetone phosphate. The acylation of dihydroxyacetone phosphate by acyl-CoA:dihydroxyacetone-phosphate acyltransferase is the initial and obligatory step in the synthesis of alkyl phospholipids.[13,14] This pathway has been demonstrated in a wide variety of tissues and organs; however, the relative importance of this route for the synthesis of diacyl phospholipids has not been clarified.[15] In the Zellweger syndrome, a lethal hereditary disease characterized by the absence of peroxisomes in liver and kidney, the level of alk-1-enyl phospholipids (plasmalogens) has been found to be less than 10% of that found in normal tissues, although no other abnormalities in cellular phospholipids were detected.[16] A severe deficiency in the activity of acyl-CoA:dihydroxyacetone-phosphate acyltransferase in liver, brain, and cultured skin fibroblasts from these patients has also been observed.[17,18] The acyl/alkyl-dihydroxyacetone reductase activity in tissues from Zellweger syndrome was almost normal, but a reduced activity of alkyl-dihydroxyacetone-phosphate synthetase (by 60%) and glycerol-3-phosphate acyltransferase (by 40%) were also seen.[18] In CHO cell mutants defective in the peroxisomal enzyme acyl-CoA:dihydroxyacetone-phosphate acyl-transferase, the content of plasmalogens is markedly reduced, this decrease being compensated by an increase in the diacyl form of phosphatidylethanolamine.[19] In rat liver homogenates incubated in the presence of physiological concentrations of glycerol-3-phosphate, dihydroxyacetone phosphate, and acyl-CoA, the acylation of dihydroxyacetone phosphate was about 7% of the total.[20] These results support the view that the initial step for the synthesis of the majority of the diacyl phospholipids is glycerol-3-phosphate, and not dihydroxyacetone phosphate. However, a glycerol-3-phosphate dehydrogenase null mutant has been obtained in BALB/cHcA mice in which the morphological, physiological, and reproducible characteristics of the mutant mice appear normal. This suggests that alternate pathways of energy metabolism and lipid synthesis might exist which obviate the function of glycerol-3-phosphate dehydrogenase. The role of glycerol-3-phosphate dehydrogenase may be assumed by glycerol-P-acyltransferase, an enzyme that acts on both glycerol-3-phosphate and dihydroxyacetone phosphate.[161]

A rapid stimulation of the *de novo* synthesis of phosphatidic acid was first shown in adrenals in response to ACTH and cyclic AMP.[21] ACTH caused a 50 to 100% increase in adrenal concentrations of phosphatidic acid, phosphatidylinositol, and phosphatidylglycerol within 1 to 15 min.[21-23] It has also been shown that in BC3H1 myocytes, insulin stimulates the *de novo* synthesis of phosphatidic acid and diacylglycerol.[24,145] Although the exact mechanism by which these signals stimulate phospholipid synthesis is not known, the results indicate that hormones not only modulate the turnover of phospholipids, but also the *de novo* synthesis.[25]

Phosphatidic acid can also be synthesized as a consequence of the phosphorylation of 1,2-diacylglycerol by diacylglycerol kinase, a widely distributed enzyme in mammalian tissues.[26] The synthesis of phosphatidic acid via diacylglycerol kinase serves to scavenge the diacylglycerol formed as a product of the turnover of membrane phospholipids, and does not contribute to the *de novo* synthesis of phospholipids. Numerous studies have shown a rapid accumulation of 1,2-diacylglycerol and the appearance of newly labeled phosphatidic acid upon cell stimulation.[4] Diacylglycerol is thought to be formed by the phosphodiesterasic hydrolysis of inositol phospholipids catalyzed by phospholipase C, yet it has now become clear that inositol phospholipids are not the only source of diacylglycerol. Thus, in many types of cells, phorbol esters have been shown to stimulate phosphatidylcholine hydrolysis (or phosphatidylethanolamine[164]) and the formation of diacylglycerol.[27-30] Moreover, diacylglycerol can also be formed from monoacylglycerol, as has been shown in BALB/C 3T3-cells stimulated with PDGF.[172] In culture fibroblasts stimulated with α-thrombin, present evidence indicates that, whereas the phosphoinositides are responsible for part of the molecules of diacylglycerol generated during the first 15 s after the addition of the signal, phosphatidylcholine contributes to the majority of the diacylglycerol generated after 5 min. The results are compatible with a phospholipase C reaction, although it is difficult to exclude the possibility that a reversal of the cholinephosphotransferase reaction is responsible for the generation of diacylglycerol (see Figure 1).[172] A phospholipase C acting on phosphatidylcholine, but not on phosphatidylinositol, has been identified in mammalian systems,[31] and a phosphatidylcholine phospholipase C that is stimulated by guanine nucleotides has been described in rat liver plasma membranes.[159] Whether these enzymes are activated by the addition of phorbol esters or hormones remains to be determined. Phosphatidic acid can also be formed as a result of the reaction of phospholipase D with phospholipids. In several systems, evidence has been provided which suggests that phospholipase D activation might be one of the early lipolytic events during the hydrolysis of phosphatidylcholine stimulated by hormones or phorbol esters.[32-34,160,163,164,170] In these systems, it is thought that the rise in diacylglycerol is mediated by phospholipase D and a phosphatidate phosphohydrolase, and not by the action of phospholipase C (see Figure 1). It is possible that both types of phospholipases might act, and that their relative importance varies with the cell type or signal used. In fact, signals such as phorbol esters might stimulate a cyclical degradation (to diacylglycerol) and resynthesis of phosphatidic acid similar to the well-known phosphatidylinositol cycle. The observation that protein kinase C is controlled by the 1,2-diacylglycerol generated during cell activation, that this diacylglycerol can be an important source of arachidonic acid[3,4] (the necessary precursor for the synthesis of leukotrienes and prostaglandins), and that phosphatidic acid stimulates cell growth[5] have led to a renewed interest in studies of phosphatidic acid synthesis by phosphorylation of diacylglycerol, since this is a mechanism of control of protein kinase C and of the levels of arachidonic acid. Moreover, it is also possible that phospholipase D might act on a pool of choline-linked phosphoglycerides different from phospholipase C. Thus, stimulation of polymorphonuclear leukocytes results in hydrolysis of 1-O-alkyl-2-acyl glycerophosphocholine to yield 1-O-alkyl-2-acyl glycerophosphatidate and 1-O-alkyl-2-acylglycerol, a process which is probably initiated by activation of a phospholipase D.[163]

Diacylglycerol kinase is distributed between soluble and particulate fractions, and in the rat liver about 50 and 30% of the initial activity is recovered in the soluble and particulate fractions, respectively.[35] Diacylglycerol kinase has been purified from pig[26] and rat[36] brain cytosol, and a polyclonal antibody raised against the pig brain-soluble enzyme precipitated the soluble, microsomal, and synaptosomal activities,[37] suggesting that the same form of enzyme is present in all three fractions. Translocation of the enzyme from cytosol to membranes has been observed in brain and liver rat homogenates incubated with phospholipase C or diacylglycerol[36] and in neutrophils stimulated with chemotactic peptide or phorbol esters.[38] Vasopressin and phenylephrine have been found to activate rat liver plasma-membrane diacylglycerol kinase.[39] Whether this effect is due to promoting the translocation of the enzyme from the cytosol to the plasma membrane or by covalent modification of the enzyme is not known. The activation by vasopressin and phenylephrine of diacylglycerol kinase could be an important factor in controlling the concentration of diacylglycerol formed as a product of the turnover of membrane phospholipids and probably of protein kinase C activity. In spontaneously hypertensive rat erythrocytes, diacylglycerol kinase has been found to be inhibited and phospholipase C to be activated.[76] This might be responsible for the accumulation of diacylglycerol[76] and the sustained activation of protein kinase C[77] found in erythrocytes from spontaneously hypertensive rats, and may play an important role in sustained vascular smooth muscle constriction. However, it is important to mention at this point that enzymatic studies using lipid substrates are complicated, due to the insoluble nature of the substrate, and it is not always obvious how the activities observed with aqueous dispersions of lipids might be related to the equivalent situation *in vivo*.[40,41] Finally, a defect in diacylglycerol kinase activity seems to be the primary cause of retinal degeneration in rdgA *Drosophila* visual mutant that shows age-dependent retinal degeneration.[165]

By virtue of its location, phosphatidic acid plays a key role in the metabolism of glycerolipids, and phosphatidate phosphohydrolase, the enzyme that converts phosphatidic acid into 1,2-diacylglycerol (Figure 1), has long been considered to be a potentially regulatory enzyme which could influence the production of both acidic and zwiterionic phospholipids as well as neutral glycerides. A variety of tissues contain both Mg^{2+}-dependent, or at least Mg^{2+}-stimulated, and Mg^{2+}-independent phosphatidate phosphohydrolase.[42-44] The Mg^{2+}-dependent enzyme is present in both the microsomal and cytosolic fraction of liver, adipose tissue, and lung, and this form of phosphatidate phosphohydrolase is thought to be the most important in glycerolipid synthesis.[42-44] Thus, studies with lung have demonstrated that when the Mg^{2+}-dependent phosphohydrolase is removed from the microsomes with salts, the microsomes lose their capacity to incorporate [^{14}C]glycerol-3-phosphate through phosphatidic acid into neutral lipids and phosphatidylcholine.[45] Addition to the membranes of the microsomal wash, or cytosolic Mg^{2+}-dependent phosphatidate phosphohydrolase, restored glycerolipid synthesis by the microsomes.[45] Furthermore, in the liver, it has been observed that changes in the soluble and microsomal activities of phosphatidate phosphohydrolase under a variety of physiological and pharmacological conditions are greater than those observed for other enzymes involved in triglyceride synthesis, and that these changes normally parallel the liver's capacity to synthesize triglycerides.[42]

Studies in liver have shown that the Mg^{2+}-dependent form of phosphatidate phosphohydrolase is activated two- to fourfold by glucocorticoids, glucagon, and cyclic AMP derivatives, and that the effect of glucocorticoids and glucagon are antagonized by insulin.[46-49] The relative proportion of phosphatidate phosphohydrolase associated with membranes also changed in response to these agents, but the magnitude of this effect was very small and probably without physiological implications. Thus, whereas glucagon and dexamethasone stimulated total phosphatidate phosphohydrolase by 172 and 422%, respectively, the amount of enzyme associated to membranes varied from $25 \pm 8\%$ in controls to $15 \pm 6\%$ in the presence of glucagon and $15 \pm 4\%$ in response to dexamethasone.[49] In hepatocytes, vasopressin, a hormone that causes Ca^{2+} mobilization, also stimulates phosphatidate phosphohydrolase and glycerolipid synthesis from added oleate.[50] In adipose tissue, agents that interact with β-adrenergic receptors control phosphatidate phosphohydrolase activity through the translocation of this enzyme from the cytosol to the endoplasmic reticulum.[51] An increase in microsomal phosphohydrolase has also been observed when livers were perfused with oleate in the presence of micromolar concentrations of dibutyryl cAMP.[52] In isolated rat hepatocytes[53] and in a cell line from human lung,[54] oleic acid promotes the activation and translocation of phosphatidate from the cytosol to membrane fractions. In contrast to the situation with glucagon and dexamethasone, the magnitude of this effect is large and probably of physiological interest. Thus, the addition of 4 mM oleate to hepatocytes increased the relative proportion of particulate phosphohydrolase from 30 to 97%.[52] The mechanism by which agents like glucagon or dexamethasone stimulate phosphatidate phosphohydrolase activity is not known, but they seem to be related to protein synthesis, since the stimulation can be prevented by inhibition of DNA transcription or protein synthesis.[46-48] The mechanism involved in enzyme translocation is also not known. Incubation of rat liver cytosol with alkaline phosphatase has been shown to stimulate phosphatidate phosphohydrolase activity by 20 to 70%.[55,56] Although suggestive, more direct evidence is necessary to conclude that phosphatidate phosphohydrolase is modulated by reversible phosphorylation.

Amphiphilic drugs which, like monodansylcadaverine, possess an ionizable amino group with a hydrophobic substituent increase the *de novo* synthesis of phosphatidic acid and diacylglycerol.[57] This effect of amphiphilic drugs is accompanied by an increased synthesis of phosphatidylinositol in a variety of cells, including lymphocytes,[58] pineal gland,[59] human and rat neutrophils,[57,60] and human and chicken embryo fibroblasts.[60] This effect on phosphatidylinositol synthesis was accompanied by a decreased synthesis of phosphatidylcholine, phosphatidylethanolamine, and triglycerides in lymphocytes[58] and neutrophils,[57,60] and by decreased synthesis of phosphatidylcholine and triglycerides, but not of phosphatidylethanolamine, in pineal gland.[59] These results are suggestive of an inhibition by amphiphilic drugs of the conversion of diacylglycerol to phosphatidylcholine and phosphatidylethanolamine, i.e., a redirection of lipid synthesis from these phospholipids to diacylglycerol, phosphatidic acid, phosphatidylinositol, and triglycerides. Amphiphilic drugs such as monodansylcadaverine are well-known inhibitors of the receptor-mediated endocytosis of many ligands.[61] Taking into account the marked effects of these drugs on lipid metabolism in a variety of cell types, it seems possible that these agents might inhibit cellular functions such as endocytosis and chemotaxis by profoundly altering phospholipid metabolism.

III. PHOSPHATIDYLCHOLINE AND PHOSPHATIDYLETHANOLAMINE

Figure 3 shows the sequence of reactions leading to the synthesis of phosphatidylcholine and phosphatidylethanolamine. The final reaction for the synthesis of both phospholipids is the transfer of phosphocholine or phosphoethanolamine from CDP-choline and CDP-ethanolamine, respectively, to a molecule of 1,2-diacylglycerol catalyzed by two separate transferases.[62,63] In addition, phosphatidylcholine can be produced by three sequential N-methylations of preexisting phosphatidylethanolamine.[64,65] From a quantitative point of view, the methylation pathway for the synthesis of phosphatidylcholine is only important in the liver (and probably in adipocytes and Leydig cells), where up to 20 to 40% of the phosphatidylcholine may be produced in this way,[66] whereas in other tissues like in the brain, it has been estimated that less than 1% of the total phosphatidylcholine originates by the methylation pathway.[67] This pathway, however, is probably the only source of choline moieties in animals.[68,69] Furthermore, several reports indicate that the methylation pathway provides species of phosphatidylcholine rich in polyunsaturated fatty acids (docosahexanoic and arachidonic acids),[70-75] and that the turnover of these polyunsaturated molecules of phosphatidylcholine is faster than that of the more saturated species.[75]

The first step in the conversion of choline to CDP-choline is the formation of phosphorylcholine by choline kinase (Figure 3). Choline kinase catalyzes the phosphorylation of choline by ATP in the presence of Mg^{2+} to generate phosphocholine and ADP.[78] Several forms of choline kinase have been detected in rat tissues.[79,80] One of these forms has been purified to homogeneity from rat kidney cytosol.[80] The purified enzyme has an M_r of 42 kDa and probably forms a dimer with an M_r of 75 to 80 kDa.[80] The 42-kDa form does not exist in extracts from rat liver or lung. In chicken liver, choline kinase has an M_r of 36 kDa, as estimated by gel filtration.[81] In rat tissues, the same enzyme seems to be responsible for both choline and ethanolamine phosphorylation.[82] cAMP-dependent phosphorylation and inactivation of the chicken liver enzyme has been reported.[81] Furthermore, choline kinase from *Saccharomyces cerevisiae* has been cloned and found to contain two putative phosphorylation sites near the N-terminal region of the encoded polypeptide.[83] Finally, choline kinase has been shown to be induced two- to fourfold in response to a variety of agents. These agents include fatty acids,[84] carbon tetrachloride,[85,86] estradiol,[87] and insulin.[88,89] More recently, elevated choline kinase activity has been detected in *ras*-transformed NIH 3T3 cells, and this might be the cause of the increased concentration of phosphocholine found in these cells.[171]

CTP:phosphocholine cytidylyltransferase catalyzes the transfer of a cytidylyl moiety from CTP to choline phosphate, the reaction product of CDP-choline and pyrophosphate (Figure 3). The synthesis of CDP-choline by this enzyme has been shown to be the rate-limiting step in phosphatidylcholine synthesis in a variety of tissues, including liver,[90] lung,[91] and granular pneumocytes.[92] Cytidylyltransferase is present in both the cytosol and membrane fractions of a variety of tissues. In rat lung, the activity of cytosolic phosphocholine cyti-

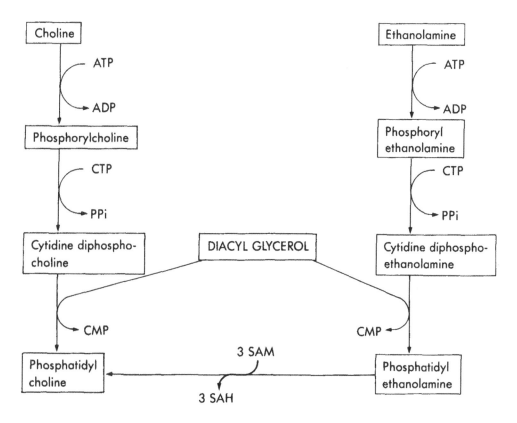

FIGURE 3. *Schematic representation of the pathway of biosynthesis of phosphatidylcholine and phosphatidylethanolamine. SAM, S-adenosyl-L-methionine; SAH, S-adenosyl-L-homocysteine.*

dylyltransferase increases by either aggregation in the presence of lipids[93] or by translocation to the endoplasmic reticulum.[91,94,104] Regulation of cytidylyltransferase activity by subcellular translocation has been described in various tissues.[95-97] Factors that induce phosphocholine cytidylyltransferase translocation include fatty acids,[90,98] phosphatidylglycerol,[99,100] phorbol esters,[101,102] choline depletion,[103] phospholipase C,[95,96,104] and thyrotropin-releasing hormone,[105] but the molecular mechanisms of this process remain to be determined. Phosphocholine cytidylyltransferase has been purified from rat liver,[166-168] and present evidence indicates that the purified enzyme is a homodimer with an M_r about 42 kDa.[168]

Incubation of rat hepatocytes for short periods with cAMP analogs has been reported to result in an inhibition of phosphatidylcholine synthesis.[106] However, incubation for longer periods resulted in a stimulation of phosphatidylcholine biosynthesis in both rat hepatocytes and fetal rat lung explants or transformed adult human lung cell lines.[106-180] These results suggest that although fetal lung and liver phosphocholine cytidylyltransferase may be regulated by reversible phosphorylation, other phenomena may occur as well during exposure of cells to cAMP analogs or conditions that increase cAMP levels. Microsomal and cytosolic cytidylyltransferase has been found to be inhibited after preincubation with MgATP.[109] In liver, activation of the cytosolic enzyme by the addition of phospholipid vesicles is inhibited by the presence of MgATP,[110] and in both liver and fetal lung, the inhibition of phosphocholine cytidylyltransferase by MgATP was inhibited by the heat-stable protein inhibitor of cAMP-dependent protein kinase. Finally, incubation of purified rat liver phosphocholine cytidylyltransferase with [τ-^{32}P]ATP and cAMP-dependent protein kinase resulted in phosphorylation of the enzyme.[162] These results indicate that conditions that favor protein phosphorylation inhibit cytidylyltransferase activity. Whether the enzyme is actually phosphorylated and/or phosphorylation is important for the translocation of the enzyme *in vivo* remains to be proven. Phorbol esters, 1-oleoyl-2-acetylglycerol, and other lipids that are well-known activators of protein kinase C have been found to stimulate phosphocholine cytidylyltransferase.[101-103] The mechanism by which these agents exert their action on phosphatidylcholine synthesis is, however, not known. In the case of 1-oleoyl-2-acetylglycerol and diacylglycerol, a direct activation of cytidylyltransferase by these lipids has been proposed.[103] In NG108-15 neuroblastoma × glioma hybrid cells, the regulation of phosphatidylcholine synthesis by two cell-permeant diacylglycerols takes place through a mechanism different from that triggered by phorbol esters. This is based on the observation that whereas in cells down-regulated for protein kinase C, the effect of additional phorbol ester on choline incorporation into phosphatidylcholine was abolished, diacylglycerol-dependent phosphatidylcholine biosynthesis was not affected.[111] Similarly, in GH_3 pituitary cells down-regulated for protein kinase C, the effect of TRH or additional phorbol ester on phosphatidylcholine synthesis was also abolished.[105] These results suggest that certain hormones may regulate phosphatidylcholine metabolism through protein kinase activation.

A regulatory role for cholinephosphotransferase, the enzyme that catalyzes the transfer of a phosphocholine moiety from CDP-choline to diacylglycerol, with phosphatidylcholine and CMP the products of the reaction (Figure 3), has also been suggested in the biosynthesis of phosphatidylcholine in the lung. This conclusion is based on the observation that CDP-choline accumulates during the development and hormonal induction of fetal rabbit lung.[112,113] This might be due to changes in the levels of diacylglycerol[113,114] and/or to the existence of different diacylglycerol pools[115,172] for the biosynthesis of phosphatidylcholine. In lung, liver, and muscle microsomes, it has been shown that endogenous diacylglycerol and diacylglycerol synthesized *de novo* by acylation of glycerol-3-phosphate exist as two different substrate pools for the synthesis of phosphatidylcholine by the choline phosphotransferase reaction.[115] It has been proposed that the synthesis of phosphatidylcholine takes place in a multienzyme complex, where the diacylglycerol newly synthesized through acylation of glycerol-3-phosphate is preferentially used by choline phosphotransferase.[115] According to this model, if the supply of CDP-choline exceeds the diacylglycerol formed *de novo*, the choline phosphotransferase reaction might become the rate-limiting step in phosphatidylcholine synthesis. Furthermore, these results suggest that the diacylglycerol formed by phosphatidylcholine degradation might not be readily available for phospholipid synthesis, which might be important in view of the role of diacylglycerol as an activator of protein kinase C.

Phosphatidylcholine can also be synthesized by three successive N-methylations of phosphatidylethanolamine with S-adenosylmethionine as the methyl donor, the mono- and dimethyl derivatives of phosphatidylethanolamine being the intermediates of the reaction (Figure 3).[64,65] In this reaction, S-adenosylmethionine is converted to S-adenosylhomocysteine, a competitive inhibitor of transmethylation reactions. The results of classical genetics studies with *Neurospora crassa* and *S. cerevisiae* mutants defective in the phospholipid methylation pathway showed that in one type of mutant, the conversion of phosphatidylethanolamine to phosphatidyl-*N*-monomethylethanolamine was greatly reduced, and the conversion of phosphatidyl-*N*-monomethylethanolamine to phosphatidyl-*N,N*-dimethylethanolamine and then to phosphatidylcholine, was deficient in another type of mutant.[116-118] These studies were interpreted as suggestive of the existence of two different methyltransferases, one converting phosphatidylethanolamine to phosphatidyl-*N*-monomethylethanolamine, and a second enzyme catalyzing the conversion of phosphatidyl-*N*-monomethylethanolamine to phosphatidyl-*N,N*-dimethylethanolamine and then to phosphatidylcholine.[116-118] More recently, the structural genes (PEM1 and PEM2) encoding the enzymes involved in yeast phospholipid methylation have been cloned.[119] The product of PME1, an enzyme of M_r 101 kDa called phosphatidylethanolamine methyltransferase, catalyzed the conversion of phosphatidylethanolamine to phosphatidyl-*N*-monomethylethanolamine, and the product of PEM2, an enzyme of M$_r$ 23 kDa called phospholipid methyltransferase, catalyzed the synthesis of phosphatidylcholine from phosphatidylethanolamine.[119] Although the product of PEM2 contained all three methylation activities, the mono- and dimethyl derivatives of phosphatidylethanolamine were used preferentially rather than phosphatidylethanolamine. In mammalian tissues, present evidence indicates that in rat liver a single enzyme with a reported M$_r$ of 25 kDa[120] or 18 kDa[121] synthesizes phosphatidylcholine from phosphatidylethanolamine. This enzyme might be the mammalian version of yeast PEM2. Whether mammalian tissues posses a second methyltransferase which corresponds to the yeast PEM1 and catalyzes only the conversion of phosphatidylethanolamine to phosphatidyl-*N*-monomethylethanolamine is not known.

In mammalian tissues, the specific activity of phospholipid methyltransferase is maximal in rat liver, adipose tissue, and Leydig cells.[120-122] It is interesting to note that in these three tissues, the addition of cAMP, cAMP analogs or conditions that elevate the cellular cAMP content (e.g., glucagon, isoproterenol, forskolin, cholera toxin, etc.) stimulate phospholipid methyltransferase,[120-126] probably by enhancing the phosphorylation state of the enzyme.[127-129] In rat liver and adipose tissue, insulin has been shown to block the effect of isoproterenol, ACTH, and glucagon on phospholipid methyltransferase,[123,125,126] probably by favoring the dephosphorylated state of the enzyme.[128,129] Interestingly, exposure of rat hepatocytes to cAMP for short periods has been shown to produce an inhibition of phosphatidylcholine synthesis by the CDP-choline pathway.[106] A similar situation was observed with conditions that elevate cellular Ca^{2+}. Thus, whereas vasopressin, angiotensin, and oxytocin have been shown to stimulate phospholipid methyltransferase,[130] probably through phosphorylation of the enzyme,[132] conditions that elevate intracellular Ca^{2+} (i.e., vasopressin and A23187) decreased the synthesis of phosphatidylcholine by the CDP-choline pathway.[133] This led to the hypothesis of a coordinate mechanism of regulation of the synthesis of phosphatidylcholine by both pathways.[134] Dietary conditions also result in the activation of one pathway and the inhibition of the alternate pathway of phosphatidylcholine biosynthesis. Thus, whereas the addition of choline to rats stimulates the CDP-choline route, this treatment inhibited the methylation pathway.[135-138] Similarly, in liver from diabetic rats, phospholipid *N*-methyltransferase is inhibited[139,140] and the CDP-choline pathway is stimulated.[139] The purpose of this coordinate mechanism of control of phosphatidylcholine synthesis in the liver might be to maintain a relatively constant ratio of phosphatidylcholine to phosphatidylethanolamine under a variety of conditions. This agrees with data obtained from the liver of rats subjected to different nutritional conditions, where the ratio of phosphatidylcholine to phosphatidylethanolamine is kept close to constant despite large variations in the amount of phosphatidylcholine.[141]

In addition to its role in phosphatidylcholine synthesis, the methylation pathway is probably the only source of choline moieties in animals.[68,69] In this respect, it is interesting to note that in rats fed a choline-deficient diet, there is a marked stimulation of phospholipid *N*-methyltransferase and an inhibition of the CDP-choline pathway.[135-138] Furthermore, several reports indicate that the methylation pathway provides species of phosphatidylcholine rich in polyunsaturated fatty acids (docosahexanoic and arachidonic acids),[70-75] and that the turnover of these polyunsaturated molecules of phosphatidylcholine is faster than that of the more saturated species.[75] This might be due both to the high content of unsaturated species in the pool of phosphatidylethanolamine[70-75] and to a certain specificity of phospholipid methyltransferase for those molecules of phosphatidylethanolamine with two or more double bonds.[142] Although there is a considerable body of evidence linking phospholipid methylation and signal transduction during a variety of biological processes, in tissues such as the brain, where it has been estimated that less than 1% of the total phosphatidylcholine originates by the methylation pathway,[67] the function of this route remains obscure and the nature and function of the resulting biochemical signals remains to be determined.[134] In the brain, the function of phospholipid methylation might be to provide a source of choline for acetylcholine synthesis as well as to contribute to the synthesis of polyunsaturated species of phosphatidylcholine.[75,143,144] A further role of phospholipid methylation might be to complete a phosphatidylethanolamine/phosphatidylcholine cycle in which phosphatidylcholine molecules generated by the transmethylation pathway are hydrolyzed to form diacylglycerol, which is then converted to phosphatidylethanolamine. The following results mentioned previously favor this hypothesis:

1.   A number of signals stimulate the phosphodiesteratic hydrolysis of phosphatidylcholine (presumably of those rich in polyunsaturated fatty acids) to yield diacylglycerol.
2.   Molecules of phosphatidylcholine rich in unsaturated fatty acids have a faster turnover than those with more saturated fatty acids.
3.   A variety of signals stimulate the conversion of phosphatidylethanolamine to phosphatidylcholine.

Phosphatidylethanolamine contains high amounts of polyunsaturated fatty acids. This specific composition is thought to be accomplished, as in the case of other phospholipids, through deacylation-reacylation reactions.[146] In addition, the direct transfer of phospholipid acyl groups to lysophospholipid acceptors (i.e., from phosphatidylcholine to lysophosphatidylethanolamine) in the absence and presence of CoA has been shown to occur in homogenates from a variety of cells.[147-151] Based on the fatty acid specificity of these transfer reactions, it has been proposed that this is an important mechanism by which phospholipids achieve their specific fatty acid composition.

Phosphatidylethanolamine can be synthesized in mammalian cells by the transfer of phosphoethanolamine from CDP-ethanolamine to diacylglycerol[62,63,152,153] and by decarboxylation of phosphatidylserine (Figures 1 and 2).[154] Phosphatidylserine decarboxylation has been reported to account for nearly all phosphatidylethanolamine synthesized in a hamster kidney cell line.[155] A similar conclusion has been drawn using CHO mutants that, although deficient in the CDP-ethanolamine pathway, had a normal cellular content of phosphatidylethanolamine[152,153] and an incorporation of ^{32}P into this lipid similar to that found in the parental cell line.[152] Phosphatidylserine biosynthesis occurs in mammals mainly by the base exchange of free serine with the polar headgroup of preexisting phosphatidylcholine or phosphatidylethanolamine, although a new pathway depending on CMP has been reported in rat liver microsomes.[156] Additional support for the conclusion that phosphatidylserine is the main source of phosphatidylethanolamine in mammalian cells comes from the studies with CHO base-exchange mutants. When grown in the absence of phosphatidylethanolamine, these mutants not only have a decreased content of phosphatidylserine, but also of phosphatidylethanolamine.[157,158]

These results indicate a complex network of enzymatic reactions linking the biosynthesis of the various cellular phospholipids. Thus, phosphatidylcholine is mainly synthesized by the CDP-choline pathway, then to be converted into phosphatidylserine by base exchange, and phosphatidylserine is converted into phosphatidylethanolamine by decarboxylation, then to be converted to phosphatidylcholine by the N-methylation pathway. Additionally, phosphatidylethanolamine can be synthesized by the CDP-ethanolamine pathway and, under certain conditions, phosphatidylethanolamine can serve as a source of phosphatidylserine by base exchange. The relative importance of each pathway might differ from cell to cell, and, additionally, different molecular species of phospholipids might use different pathways in the same cell.

REFERENCES

1. **Dawidowicz, E. A.**, Dynamics of membrane lipid metabolism and turnover, *Annu. Rev. Biochem.*, 56, 43, 1987.
2. **Hokin, L. E.**, Receptors and phosphoinositide-generated second messengers, *Annu. Rev. Biochem.*, 54, 205, 1985.
3. **Nishizuka, Y.**, The role of protein kinase C in cell surface signal transduction and tumor promotion, *Nature*, 308, 693, 1984.
4. **Berridge, M. J.**, Inositol triphosphate and diacylglycerol: two interacting second messengers, *Annu. Rev. Biochem.*, 56, 159, 1987.
5. **Moolenar, W. H., Kruijer, W., Tilly, B. C., Verlaan, I., Bierman, A. J., and de Laat, S. W.**, Growth factor-like action of phosphatidic acid, *Nature*, 310, 644, 1984.
6. **Hannun, Y., Loomis, C., Merrill, A., and Bell, R.**, Sphingosine inhibition of protein kinase C activity and of phorbol dibutyrate binding in vitro and in human platelets, *J. Biol. Chem.*, 261, 12604, 1986.
7. **Bremer, E. G., Hakomori, S., Bowen-Pope, D. F., Raines, E., and Ross, R.**, Ganglioside-mediated modulation of cell growth. Growth factor binding, and receptor phosphorylation, *J. Biol. Chem.*, 259, 6818, 1984.
8. **Alemany, S., Mato, J. M., and Stralfors, P.**, Phospho-dephospho-control by insulin is mimicked by a phospho-oligosaccharide in adipocytes, *Nature*, 330, 77, 1987.
9. **Winslon, C. M. and Lee, M. L.**, *New Horizons in Platelet Activating Factor Research*, John Wiley & Sons, New York, 1987.
10. **Hübscher, G.**, Glyceride metabolism, in *Lipid Metabolism*, Wakil, S. J., Ed., Academic Press, New York, 1970, 279.
11. **Kennedy, E. P.**, Synthesis of phosphatides in isolated mitochondria, *J. Biol. Chem.*, 201, 399, 1953.
12. **Kornberg, A. and Procer, W. E., Jr.**, Enzymatic esterification of α-glycerophosphate by long chain fatty acids, *J. Biol. Chem.*, 204, 345, 1953.
13. **Hajra, A. K.**, Acyl dyhydroxyacetone phosphate: precursor of alkyl ethers, *Biochem. Biophys. Res. Commun.*, 39, 1037, 1970.
14. **Wykle, R. L., Piantadosi, C., and Snyder, F.**, The role of acyldihydroxyacetone phosphate, NADH, and NADPH in the biosynthesis of O-alkyl glycerolipids by microsomal enzymes of Ehrlich ascites tumor, *J. Biol. Chem.*, 247, 2944, 1972.
15. **Hajra, A. K.**, Biosynthesis of glycerolipids via acyldihidroxyacetone phosphate, *Biochem. Soc. Trans.*, 5, 34, 1977.
16. **Heymans, H. S. A., Schutgens, R. B. H., Tan, R., van den Bosch, H., and Borst, P.**, Severe plasmalogen deficiency in tissues of infant without peroxisomes (Zellweger syndrome), *Nature*, 306, 69, 1983.
17. **Schutgen, R. B. H., Romeyn, G. S., Wanders, R. J. A., van den Bosch, H., Schrakamp, G., and Heymans, H. S. A.**, Deficiency of acyl-CoA: dihydroxyacetone-phosphate acyltransferase in patients with Zellweger (cerebro-hepato-renal) syndrome, *Biochem. Biophys. Res. Commun.*, 120, 179, 1984.
18. **Datta, N. S., Wilson, G. N., and Hajre, A. K.**, Deficiency of enzymes catalyzing the biosynthesis of glycerol-ether lipids in Zellweger syndrome, *N. Engl. J. Med.*, 311, 1080, 1984.
19. **Zoeller, R. A. and Raetz, C. R. H.**, Isolation of animal cell mutants deficient in plasmalogen biosynthesis and peroxisome assembly, *Proc. Natl. Acad. Sci. U.S.A.*, 83, 5170, 1986.
20. **Declercq, P. E., Haagsman, H. P., van Veldhoven, P. V., Debeer, L. J., van Golde, L. M. G., and Mannaerts, G. P.**, Rat liver dihydroxyacetone-phosphate acyltransferases and their contribution to glycerolipid synthesis, *J. Biol. Chem.*, 259, 9064, 1984.
21. **Farese, R. V., Sabir, M. A., and Larson, R. E.**, On the mechanism whereby ACTH and cyclic AMP increase adrenal polyphosphoinositides. Rapid stimulation of the synthesis of phosphatidic acid and derivatives of CDP-diacylglycerol, *J. Biol. Chem.*, 255, 7232, 1980.
22. **Farese, R. V., Sabir, M. A., and Vandor, S. L.**, Adrenocorticotropin acutely increases adrenal polyphosphoinositides, *J. Biol. Chem.*, 254, 6842, 1979.
23. **Farese, R. V., Sabir, M. A., and Larson, R. E.**, Effects of adrenocorticotropin and cycloheximide on adrenal diglyceride kinase, *Biochemistry*, 20, 6047, 1981.
24. **Farese, R. V., Barnes, D. E., Davis, J. S., Standaert, M. L., and Pollet, R. J.**, Effects of insulin and protein synthesis inhibitors on phospholipid metabolism, diacylglycerol levels, and pyruvate dehydrogenase activity in BC3H-1 cultured myocytes, *J. Biol. Chem.*, 259, 7094, 1984.
25. **Farese, R. V.**, *De novo* phospholipid synthesis as an intercellular mediator system, in *Phospholipids and Cellular Regulation*, Kuo, J. R., Ed., CRC Press, Boca Raton, FL, 1985, chap. 7.
26. **Kanoh, H., Kondoh, H., and Ono, T.**, Diacylglycerol kinase from pig brain, *J. Biol. Chem.*, 258, 1767, 1983.
27. **Besterman, J. M., Duronia, V., and Cuatrecasas, P.**, Rapid formation of diacylglycerol from phosphatidylcholine: a pathway for generation of second messenger, *Proc. Natl. Acad. Sci. U.S.A.*, 83, 6785, 1986.

28. **Daniel, L. W., Waite, M., and Wykle, R. L.,** A novel mechanism of diglyceride formation, *J. Biol. Chem.,* 261, 9128, 1986.
29. **Rosoff, P., Savage, N., and Dinarello, C. A.,** Interleukin-1 stimulates diacylglycerol production in T lymphocytes by a novel mechanism, *Cell,* 54, 73, 1988.
30. **Daniel, L. W., Wilcox, R. W., and Etkin, L. A.,** Diglyceride formation from phosphatidylcholine: a novel mechanism of signal transduction, *Adv. Biosci.,* 66, 225, 1987.
31. **Wolf, R. A. and Gross, R. W.,** Identification of neural active phospholipase C which hydrolyses choline glycerophospholipids and plasmalogen selectively, *J. Biol. Chem.,* 260, 7295, 1985.
32. **Liscovitch, M., Blusztajn, J. K., Freese, A., and Wurtman, R. J.,** Stimulation of choline release from NG 108-15 cells by 12-O-tetradecanoylphorbol 13-acetate, *Biochem. J.,* 241, 81, 1987.
33. **Bocckino, S. B., Blackmore, P. F., Wilson, P. B., and Exton, J. H.,** Phosphatidate accumulation in hormone-treated hepatocytes via a phospholipase D mechanism, *J. Biol. Chem.,* 262, 15309, 1987.
34. **Cabot, M. C., Welsh, C. J., Cao, H., and Chabbott, H.,** The phosphatidylcholine pathway of diacylglycerol formation stimulated by phorbol diesters occurs via phospholipase D activation, *FEBS Lett.,* 233, 153, 1988.
35. **Kanoh, H. and Akesson, B.,** Properties of microsomal and soluble diacylglycerol kinase in rat liver, *Eur. J. Biochem.,* 85, 225, 1978.
36. **Besterman, J. M., Pollenz, R. S., Booker, E. L., and Cuatrecasas, P.,** Diacylglycerol-induced translocation of diacylglycerol kinase: use of affinity-purified enzyme in a reconstitution system, *Proc. Natl. Acad. Sci. U.S.A.,* 83, 9378, 1986.
37. **Kanoh, H., Iwata, T., Ono, T., and Suzuki, T.,** Immunological characterization of sn-1,2-diacylglycerol and sn-2-monoacylglycerol kinase from pig brain, *J. Biol. Chem.,* 261, 5597, 1986.
38. **Ishitoya, J., Yamakawa, A., and Takenawa, T.,** Translocation of diacylglycerol kinase in response to chemotactic peptide and phorbol ester in neutrophils, *Biochem. Biophys. Res. Commun.,* 144, 1025, 1987.
39. **Rider, M. H. and Baquet, A.,** Activation of rat liver plasma-membrane diacylglycerol kinase by vasopressin and pheylephrine, *Biochem. J.,* 255, 923, 1988.
40. **Gatt, S. and Barenholtz, Y.,** Enzymes of complex lipid metabolism, *Annu. Rev. Biochem.,* 42, 61, 1973.
41. **Brindley, D. N. and White, D. A.,** Difficulties encountered in interpreting the kinetics of enzyme reactions involving lipid substrates, *Biochem. Soc. Trans.,* 2, 44, 1977.
42. **Brindley, D. N.,** Phosphatidate phosphohydrolase activity in the liver, in *Phosphatidate Phosphohydrolase,* Vol. 1, Brindley, D. N., Ed., CRC Press, Boca Raton, FL, 1988, chap. 2.
43. **Saggerson, E. D.,** Phosphatidate phosphohydrolase activity in adipose tissue, in *Phosphatidate Phosphohydrolase,* Vol. 1, Brindley, D. N., Ed., CRC Press, Boca Raton, FL, 1988, chap. 3.
44. **Possmayer, F.,** Pulmonary phosphatidate phosphohydrolase and its relation to the surfactant system of the lung, in *Phosphatidate Phosphohydrolase,* Vol. 2, Brindley, D. N., Ed., CRC Press, Boca Raton, FL, 1988, chap. 5.
45. **Walton, P. A. and Possmayer, F.,** The role of Mg^{2+}-dependent phosphatidate phosphohydrolase in pulmonary glycerolipid biosynthesis, *Biochim. Biophys. Acta,* 796, 364, 1984.
46. **Lehtonen, M. A., Pollard, A. D., Jennings, R. J., and Brindley, D. N.,** Hormonal regulation of hepatic soluble phosphatidate phosphohydrolase, *FEBS Lett.,* 99, 162, 1979.
47. **Jennings, R. J., Lawson, N., Fears, R., and Brindley, D. N.,** Stimulation of the activities of phosphatidate phosphohydrolase and tyrosine aminotransferase in rat hepatocytes by glucocorticoids, *FEBS Lett.,* 113, 119, 1981.
48. **Pittner, R. A., Fears, R., and Brindley, D. N.,** Effects of cyclic AMP, glucocorticoids and insulin on the activities of phosphatidate phosphohydrolase, tyrosine aminotransferase and glycerol kinase in isolated rat hepatocytes in relation to the control of triacylglycerol synthesis and gluconeogenesis, *Biochem. J.,* 225, 455, 1985.
49. **Pittner, R. A., Fears, R., and Brindley, D. N.,** Interactions of insulin, glucagon and dexamethasone in controlling the activity of glycerol phosphate acyltransferase and the activity and subcellular distribution of phosphatidate phosphohydrolase in cultured rat hepatocytes, *Biochem. J.,* 230, 525, 1985.
50. **Pollard, A. D. and Brindley, D. N.,** Effects of vasopressin and corticosterone on fatty acid metabolism and on the activities of glycerol phosphate acyltransferase and phosphatidate phosphohydrolase in rat hepatocytes, *Biochem. J.,* 217, 461, 1984.
51. **Moller, F., Wong, K. H., and Green, P.,** Control of fat cell phosphatidate phosphohydrolase by lipolytic agents, *Can. J. Biochem.,* 59, 9, 1981.
52. **Soler-Argilaga, C., Russell, R. L., and Heimberg, M.,** Enzymatic aspects of the reduction of microsomal glycerolipid biosynthesis after perfusion of the liver with dibutyryl adenosine-3′,5′-monophosphate, *Arch. Biochem. Biophys.,* 190, 367, 1978.
53. **Cascales, C., Mangiapane, E. H., and Brindley, D. N.,** Oleic acid promotes the activation and translocation of phosphatidate phosphohydrolase from the cytosol to particulate fractions of isolated rat hepatocytes, *Biochem. J.,* 219, 911, 1984.

54. **Walton, P. A. and Possmayer, F.**, Translocation of Mg^{2+}-dependent phosphatidate phosphohydrolase between cytosol and endoplasmic reticulum in a permanent cell line from human lung, *Biochem. Cell Biol.*, 64, 1135, 1986.

55. **Berglund, L., Björkhem, I., and Einarsson, K.**, Apparent phosphorylation-dephosphorylation of soluble phosphatidic acid phosphatase in rat liver, *Biochem. Biophys. Res. Commun.*, 105, 288, 1982.

56. **Butterwith, S. C., Martin, A., and Brindley, D. N.**, Can phosphorylation of phosphatidate phosphohydrolase by a cyclic AMP-dependent mechanism regulate its activity and subcellular distribution and control hepatic glycerolipid synthesis?, *Bicohem. J.*, 222, 487, 1984.

57. **Garcia Gil, M., van Lookeren Campagne, M., Esbrit, P., Navarro, F., and Mato, J. M.**, Effect of monodansylcadaverine on the synthesis of phosphatidylinositol by rabbit neutrophils, *Biochim. Biophys. Acta*, 794, 234, 1984.

58. **Allan, O. and Michell, R. H.**, Enhanced synthesis *de novo* of phosphatidylinositol in lymphocytes treated with cationic drugs, *Biochem. J.*, 148, 471, 1975.

59. **Eichberg, J., Gates, J., and Hauser, G.**, The mechanism of modification by propanolol of the metabolism of phosphatidyl-CMP (CDP-diacylglycerol) and other lipids in the rat pineal gland, *Biochim. Biophys. Acta*, 573, 90, 1979.

60. **Mato, J. M., Pencev, D., Vasanthakumar, G., Schiffmann, E., and Pastan, I.**, Inhibitors of endocytosis perturb phospholipid metabolism in rabbit neutrophils and other cells, *Proc. Natl. Acad. Sci. U.S.A.*, 80, 1929, 1983.

61. **Schlegel, R., Dickson, R. B., Willingham, M. C., and Pastan, I.**, Amantadine and dansylcadaverine inhibit vesicular stomatitis virus uptake and receptor-mediated endocytosis, *Proc. Natl. Acad. Sci. U.S.A.*, 79, 2291, 1982.

62. **Kennedy, E. P. and Weiss, S. B.**, The function of cytidine coenzymes in the biosynthesis of phospholipids, *J. Biol. Chem.*, 222, 193, 1956.

63. **Polokoff, M. A., Wing, D. C., and Raetz, C. R. H.**, Isolation of somatic cell mutants defective in the biosynthesis of phosphatidylethanolamine, *J. Biol. Chem.*, 256, 7687, 1981.

64. **Bremer, J. and Greenberg, D. M.**, Biosynthesis of choline in vitro, *Biochim. Biophys. Acta*, 37, 173, 1960.

65. **Bremer, J. and Greenberg, D. M.**, Methyl transferring enzyme system of microsomes in the biosynthesis of lecithin (phosphatidylcholine), *Biochim. Biophys. Acta*, 46, 205, 1961.

66. **Sundler, R. and Akesson, B.**, Regulation of phospholipid biosynthesis in isolated rat hepatocytes, *J. Biol. Chem.*, 250, 3359, 1975.

67. **Percy, A. K., Moore, J. F., and Waechter, C. J.**, Properties of particulate and detergent-solubilized phospholipid N-methyltransferase activity from human calf brain, *J. Neurochem.*, 38, 1404, 1982.

68. **Gibson, K. D., Wilson, J. D., and Udenfriend, S.**, The enzymatic conversion of phospholipid ethanolamine to phospholipid choline in rat liver, *J. Biol. Chem.*, 236, 673, 1961.

69. **Wise, E. M. and Elwyn, D.**, Rates of reactions involved in phosphatide synthesis in liver and small intestine of intact rats, *J. Biol. Chem.*, 240, 1537, 1965.

70. **Arvidson, G. A. E.**, Biosynthesis of phosphatidylcholines in rat liver, *Eur. J. Biochem.*, 5, 415, 1968.

71. **Kanoh, H.**, Biosynthesis of molecular species of phosphatidylcholine and phosphatidylethanolamine from radioactive precursors in rat liver slices, *Biochim. Biophys. Acta*, 176, 756, 1969.

72. **Trewhella, M. A. and Collins, F. D.**, Pathways of phosphatidylcholine biosynthesis in rat liver, *Biochim. Biophys. Acta*, 296, 51, 1973.

73. **Le Kim, D., Betzing, H., and Stoffel, W.**, Studies *in vivo* and *in vitro* on the methylation of phosphatidyl-N,N-dimethylethanolamine to phosphatidylcholine in rat liver, *Hoppe-Seyler's Z. Physiol. Chem.*, 354, 437, 1973.

74. **Tacconi, M. and Wurtman, R. J.**, Phosphatidylcholine produced in rat synaptosomes by N-methylation is enriched in polyunsaturated fatty acids, *Proc. Natl. Acad. Sci. U.S.A.*, 82, 4828, 1985.

75. **Lakher, M. B. and Wurtman, R. J.**, Molecular composition of the phosphatidylcholines produced by the phospholipid methylation pathway in rat brain, *Biochem. J.*, 244, 325, 1987.

76. **Kato, H. and Takenawa, T.**, Phospholipase C activation and diacylglycerol kinase inactivation lead to an increase in diacylglycerol content in spontaneously hypertensive rat, *Biochem. Biophys. Res. Commun.*, 146, 1419, 1917.

77. **Takaori, K., Itoh, S., Kanayama, Y., and Takeda, T.**, Protein kinase C from spontaneously hypertensive rats (SHR) and normotensive wistar Kyoto rats (WKY), *Biochem. Biophys. Res. Commun.*, 141, 769, 1986.

78. **Kennedy, E. P.**, Metabolism of lipids, *Annu. Rev. Biochem.*, 26, 119, 1957.

79. **Brophy, P. J., Choy, P. C., Toone, J. R., and Vance, D. E.**, Choline kinase and ethanolamine kinase are separate, soluble enzymes in rat liver, *Eur. J. Biochem.*, 78, 491, 1977.

80. **Ishidate, K., Nakagomi, K., and Nakazawa, Y.**, Complete purification of choline kinase from rat kidney and preparation of rabbit antibody against rat kidney choline kinase, *J. Biol. Chem.*, 259, 14706, 1984.

81. **Kulkarni, G. R. and Murthy, S. K.**, Purification of choline-ethanolamine kinase from chicken liver and its regulation by cAMP, *Indian J. Biochem. Biophys.*, 23, 90, 1986.

82. **Ishidate, K., Furusawa, K., and Nakazawa, Y.,** Complete co-purification of choline kinase and etha-nolamine from rat kidney and immunological evidence for both kinase activities residing on the same enzyme protein(s) in rat tissues, *Biochim. Biophys. Acta,* 836, 119, 1985.

83. **Hosaka, K., Kodaki, T., and Yamashita, S.,** Cloning and characterization of the yeast *CKI* gene and its expression in *Escherichia coli* (Abstr.), in 29th Int. Conf. Biochemistry of Lipids, Tokyo, September 19 to 22, 1988.

84. **Infante, J. P. and Kinsella, J. E.,** Control of phosphatidylcholine synthesis and the regulatory role of choline kinase in rat liver, *Biochem. J.,* 176, 631, 1978.

85. **Tadokoro, K., Ishidate, K., and Nakazawa, Y.,** Evidence for the existence of isozymes of choline kinase and their selective inductino in 3-methylcholanthrene or carbon tetrachloride-treated rat liver, *Biochim. Biophys. Acta,* 835, 501, 1985.

86. **Ishidate, K., Enosawa, S., and Nakazawa, Y.,** Actinomycin D-sensitive inductin of choline kinase by carbon tetrachloride intoxication in rat liver, *Biochem. Biophys. Res. Commun.,* 111, 683, 1983.

87. **Kulkarni, G. R. and Murphy, S. K.,** Induction of choline-ethanolamine kinase in chicken liver by 17-β-estradiol, *Indian J. Biochem. Biophys.,* 23, 254, 1986.

88. **Oka, T. and Perry, J. W.,** Glucocorticoid stimulation of choline kinase activity during the development of mouse mammary gland, *Dev. Biol.,* 68, 311, 1979.

89. **Ulane, R. E. and Ulane, M. M.,** The effects of insulin on choline kinase activity in cultured rat liver cells, *Life Sci.,* 26, 2143, 1980.

90. **Pelech, S. L., Pritchard, P. H., Bindley, D. N., and Vance, D. E.,** Fatty acids promote translocation of CTP:phosphocholine cytidylyltransferase to the endoplasmic reticulum and stimulate rat hepatic phos-phatidylcholine synthesis, *J. Biol. Chem.,* 258, 6782, 1983.

91. **Weinhold, P. A., Rounsifer, M. E., Williams, S. E., Brubaers, P. G., and Feldman, D. A.,** CTP:phosphocholine cytidylyltransferase in rat lung: the effect of free fatty acids on translocation of activity between microsomes and cytosol, *J. Biol. Chem.,* 259, 10315, 1984.

92. **Post, M., Batenburg, J. J., van Golde, L. M. G., and Smith, B. T.,** The rate-limiting reaction in phosphatidylcholine synthesis by alveolar type II cells isolated from fetal rat lung, *Biochim. Biophys. Acta,* 795, 558, 1984.

93. **Feldman, D. A., Kovac, C. R., Dranginis, P. L., and Weinhold, P. A.,** The role of phosphatidylglycerol in the activation of CTP:phosphocholine cytidylyltransferase from rat lung, *J. Biol. Chem.,* 253, 4980, 1978.

94. **Weinhold, P. A., Feldman, D. A., Quade, M. M., Miller, J. C., and Brooks, R. L.,** Evidence for a regulatory role of CTP:choline phosphate cytidylyl transferase in the synthesis of phosphatidylcholine in fetal lung following premature birth, *Biochim. Biophys. Acta,* 665, 134, 1981.

95. **Sleight, R. and Kent C.,** Regulation of phosphatidylcholine biosynthesis in mammalian cells. I, *J. Biol. Chem.,* 258, 824, 1983.

96. **Sleight, R. and Kent, C.,** Regulation of phosphatidylcholine biosynthesis in mammalian cells. II, *J. Biol. Chem.,* 258, 831, 1983.

97. **Sleight, R. and Kent, C.,** Regulation of phosphatidylcholine biosynthesis in mammalian cells. III, *J. Biol. Chem.,* 258, 836, 1983.

98. **Chander, A. and Fisher, A. B.,** Choline-phosphate cytidylyltransferase activity and phosphatidylcholine synthesis in rat granular pneumocytes are increased with exogenous fatty acids, *Biochim. Biophys. Acta,* 958, 343, 1988.

99. **Gilfilan, A. M., Smart, D. A., and Rooney, S. A.,** Phosphatidylglycerol stimulates cholinephosphate cytidylyltransferase activity and phosphatidylcholine synthesis in type II pneumocytes, *Biochim. Biophys. Acta,* 835, 141, 1985.

100. **Rosenberg, I. L., Smart, D. A., Gilfillan, A. M., and Rooney, S. A.,** Effect of 1-oleoyl-2-acetylglycerol and other lipids on phosphatidylcholine synthesis and cholinephosphate cytidylyltransferase activity in cultured type II pneumocytes, *Biochim. Biophys. Acta,* 921, 473, 1987.

101. **Hill, S. A., Mcmurray, W. C., and Sanwall, B. D.,** Regulation of phosphatidylcholine synthesis and the activity of CTP:cholinephosphate cytidylyltransferase in myoblasts by 12-O-tetradecanoylphorbol-13-acetate, *Can. J. Biochem.,* 62, 369, 1984.

102. **Pelech, S. L., Paddon, H. B., and Vance, D. E.,** Phorbol esters stimulate phosphatidylcholine biosynthesis by translocation of CTP:phosphocholine cytidylyltransferase from cytosol to microsomes, *Biochim. Biophys. Acta,* 795, 447, 1984.

103. **Tesan, M., Anceschi, M. M., Bleasdale, J. E.,** Regulation of CTP:cytidylyltransferase activity in type II pneumocytes, *Biochem. J.,* 232, 705, 1985.

104. **Terce, F., Record, M., Ribbes, G., Chap, H., and Douste-Blazy, L.,** Intracellular processing of cyti-dyltransferase in Krebs II cells during stimulation of phosphatidylcholine synthesis, *J. Biol. Chem.,* 263, 3142, 1988.

105. **Kolesnick, R. N.,** Thyrotropin-releasing hormone and phorbol esters induce phosphatidylcholine synthesis in GH₃ pituitary cells, *J. Biol. Chem.,* 262, 14525, 1987.

106. **Pelech, S. L., Pritchard, P. H., and Vance, D. E.,** Prolonged effects of cyclic AMP analogues on phosphatidylcholine biosynthesis in cultured rat hepatocytes, *Biochim. Biophys. Acta,* 713, 260, 1982.

107. **Gross, I. and Rooney, S. A.,** Aminophylline stimulates the incorporation of choline into phospholipids in explants of fetal rat lung in organ culture, *Biochim. Biophys. Acta,* 488, 263, 1977.

108. **Niles, R. M. and Makarski, J. S.,** Regulation of phosphatidylcholine metabolism by cyclic AMP in a model alveolar type 2 cell line, *J. Biol. Chem.,* 254, 4324, 1979.

109. **Radika, K. and Possmayer, F.,** Inhibition of foetal pulmonary choline-phosphate cytidylyltransferase under conditions favouring protein phosphorylation, *Biochem. J.,* 232, 833, 1985.

110. **Pelech, S. L. and Vance, D. E.,** Regulation of rat liver cytosolic CTP:phosphocholine cytidylyltransferase by phosphorylation and dephosphorylation, *J. Biol. Chem.,* 257, 14198, 1982.

111. **Liscovitch, M., Slack, B., Blusztajn, J. K., and Wurtman, R. J.,** Differential regulation of phosphatidylcholine biosynthesis by 12-O-tetradecanoylphorbol-13-acetate and diacylglycerol in NG108-15 neuroblastoma × glioma hybrid cells, *J. Biol. Chem.,* 262, 17487, 1987.

112. **Tokmakjian, S., Haines, D. S. M., and Possmayer, F.,** Pulmonary phosphatidylcholine biosynthesis alterations in the pool sizes of choline and choline derivatives in rabbit fetal lung during development, *Biochim. Biophys. Acta,* 663, 557, 1981.

113. **Possmayer, F.,** Pulmonary phosphatidate phosphohydrolase and its relation to the surfactant system of the lung, in *Phosphatidate Phosphohydrolase,* Vol. 2, Brindley, D. N., Ed., CRC Press, Boca Raton, FL, 1988, 39.

114. **Ide, H. and Weinhold, P. A.,** Cholinephosphotransferase in rat lung: in vitro formation of dipalmitoyl phosphatidylcholine and general lack of specificity using endogenously generated diacylglycerol, *J. Biol. Chem.,* 257, 14926, 1981.

115. **Rüstow, B. and Kunze, D.,** Further evidence for the existence of different diacylglycerol pools of the phosphatidylcholine synthesis in microsomes, *Biochim. Biophys. Acta,* 921, 552, 1987.

116. **Scarborough, G. A. and Nyc, J. F.,** Properties of a phosphatidylmonomethylethanolamine N-methyltransferase from *Neurospora crassa, Biochim. Biophys. Acta,* 146, 11, 1967.

117. **Yamashita, S. and Oshima, A.,** Regulation of the phosphatidylethanolamine methylation pathway in *Saccharomyces cerevisiae, Eur. J. Biochem.,* 104, 611, 1980.

118. **Yamashita, S., Oshima, A., Nikawa, J., and Hosaka, K.,** Regulation of phosphatidylethanolamine methyltransferase level by *myo*-inositol in *Saccharomyces cerevisiae, Eur. J. Biochem.,* 128, 5859, 1982.

119. **Kodaki, T. and Yamashita, S.,** Yeast phosphatidylethanolamine methylation pathway, cloning and characterization of two distinct methyltransferase genes, *J. Biol. Chem.,* 262, 15428, 1987.

120. **Castaño, J. G., Alemany, S., Nieto, A., and Mato, J. M.,** Activation of phospholipid methyltransferase by glucagon in rat hepatocytes, *J. Biol. Chem.,* 255, 9041, 1980.

121. **Nieto, A. and Catt, K. J.,** Hormonal activation of phospholipid methyltransferase in Leydig cell, *Endocrinology,* 113, 254, 1983.

122. **Kelly, K. L.,** Stimulation of adipocyte phospholipid methyltransferase activity by phorbol 12-myristate 13-acetate, *Biochem. J.,* 241, 917, 1987.

123. **Kraus-Friedman, N. and Zimniak, P.,** Glucagon and epinephrine-stimulated phospholipid methylation in hepatic microsomes, *Life Sci.,* 28, 1483, 1981.

124. **Schüller, A., Moscat, J., Diez, E., Fernandez-Checa, J. G., Gavilanes, F., and Municio, A. M.,** Functional properties of isolated hepatocytes from ethanol-treated rat liver, *Hepatology,* 5, 36, 1985.

125. **Marin-Cao, D., Alvarez Chiva, V., and Mato, J. M.,** Beta-adrenergic control of phosphatidylcholine synthesis by transmethylation in hepatocytes from juvenile, adult and adrenalectomized rats, *Biochem. J.,* 216, 675, 1983.

126. **Kelly, K. L., Wong, E. H. A., and Jarett, L.,** Adrenocorticotropic stimulation and insulin inhibition of adipocyte phospholipid methylation, *J. Biol. Chem.,* 260, 3640, 1985.

127. **Varela, I., Merida, I., Villalba, M., Vivanco, F., and Mato, J. M.,** Phospholipid methyltransferase phosphorylation by intact hepatocytes: effect of glucagon, *Biochem. Biophys. Res. Commun.,* 131, 477, 1985.

128. **Merida, I. and Mato, J. M.,** Inhibition by insulin of glucagon-dependent phospholipid methyltransferase phosphorylation in rat hepatocytes, *Biochim. Biophys. Acta,* 928, 92, 1987.

129. **Kelly, K. L., Merida, I., Wong, E. H. A., DiCenzo, D., and Mato, J. M.,** A phospho-oligosaccharide mimics the effect of insulin on isoproterenol-dependent phosphorylation of phospholipid methyltransferase in isolated rat adipocytes, *J. Biol. Chem.,* 262, 15285, 1987.

130. **Alemany, S., Varela, I., and Mato, J. M.,** Stimulation by vasopressin and angiotensin of phospholipid methyltransferase in isolated rat heptocytes, *FEBS Lett.,* 135, 111, 1981.

131. **Tsao, F. H. C.,** Reversibility of cholinephosphotransferase in lung microsomes, *Lipids,* 21, 498, 1986.

132. **Alemany, S., Varela, I., Harper, J. F., and Mato, J. M.,** Calmodulin regulation of phospholipid and fatty acid methylation by rat liver microsomes, *J. Biol. Chem.,* 257, 9249, 1982.

133. **Alemany, S., Varela, I., and Mato, J. M.,** Inhibition of phosphatidylcholine synthesis by vasopressin and angiotensin in rat hepatocytes, *Biochem. J.,* 208, 453, 1982.

134. **Mato, J. M. and Alemany, S.,** What is the function of phospholipid N-methylation?, *Biochem. J.*, 213, 1, 1983.

135. **Lombardi, B., Pami, F., Schlunk, F. F., and Shi-hau, C.,** Labeling of liver and plasma lecithins after injection of 1-2-(^{14})C-2-dimethylaminoethanol and (^{14})C-L-methionine-methyl to choline deficient rats, *Lipids*, 4, 67, 1969.

136. **Thompson, A., Macdonald, G., and Mookerje, S.,** Metabolism of phosphorylcholine and lecithin in normal and choline-deficient rats, *Biochim. Biophys. Acta*, 176, 306, 1969.

137. **Glenn, J. L. and Austin, W.,** The conversion of phosphatidylethanolamines to lecithins in normal and choline deficient rats, *Biochim. Biophys. Acta*, 231, 153, 1971.

138. **Schneider, W. J. and Vance, D. E.,** Effect of choline deficiency on the enzymes that synthesize phosphatidylcholine and phosphatidylethanolamine in rat liver, *Eur. J. Biochem.*, 85, 181, 1978.

139. **Hoffman, D. R., Haning, J. A., and Cornatzer, W. E.,** Effect of alloxan diabetes on phosphatidylcholine biosynthetic enzymes, *Proc. Soc. Exp. Biol. Med.*, 167, 143, 1981.

140. **Cabrero, C., Merida, I., Ortiz, P., Varela, I., and Mato, J. M.,** Effects of alloxan on S-adenosylmethionine metabolism in rats, *Biochem. Pharmacol.*, 35, 2261, 1986.

141. **Tijburg, L. B. M., Houweling, M., Geelen, M. J. H., and van Golde, L. M. G.,** Effects of dietary conditions on the pool sizes of precursors of phosphatidylcholine and phosphatidylethanolamine synthesis in rat liver, *Biochim. Biophys. Acta*, 959, 1, 1988.

142. **Ridgway, N. D. and Vance, D. E.,** Specificity of rat hepatic phosphatidylethanolamine N-methyltransferase for molecular species of diacyl phosphatidylethanolamine, *J. Biol. Chem.*, 263, 16856, 1988.

143. **Hirata, F. and Axelrod, J.,** Phospholipid methylation and biological signal transduction, *Science*, 209, 1082, 1980.

144. **Crews, F. T.,** Phospholipid methylation and membrane function, in *Phospholipids and Cellular Regulation*, Vol. 1, Kuo, J. F., Ed., CRC Press, Boca Raton, FL, 1985, 131.

145. **Farese, R. V., Cooper, D. R., Konda, T. S., Nair, G., Standaert, M. L., Davis, J. S., and Pollet, R. J.,** Mechanisms whereby insulin increases diacylglycerol in BC3H-1 myocytes, *Biochem. J.*, 256, 175, 1988.

146. **Irvine, R.,** How is the level of free arachidonic acid controlled in mammalian cells?, *Biochem. J.*, 204, 3, 1982.

147. **Kramer, R. and Deykin, D.,** Arachidonoyl transacylase in human platelets, *J. Biol. Chem.*, 258, 13806, 1983.

148. **Reddy, P. V. and Schmid, H. H. O.,** Selectivity of acyl transfer between phospholipids: arachidonoyl transacylase in dog heart membranes, *Biochem. Biophys. Res. Commun.*, 129, 381, 1985.

149. **Reddy, P. V. and Schmid, H. H. O.,** Coenzyme A-dependent and independent acyl transfer between dog heart microsomal phospholipids, *Biochim. Biophys. Acta*, 879, 369, 1986.

150. **Sugiura, T., Masuzawa, Y., Nakagawa, Y., and Waku, K.,** Transacylation of lyso platelet-activating factor and other lysophospholipids by macrophage microsomes, *J. Biol. Chem.*, 262, 1199, 1987.

151. **Reddy, P. V. and Schmid, H. H. O.,** Acylation of dog heart lysophosphatidylserine by transacylase activity, *Biochim. Biophys. Acta*, 922, 379, 1987.

152. **Polokoff, M. A., Wing, D. C., Raetz, C. R. H.,** Isolation of somatic cell mutants defective in the biosynthesis of phosphatidylethanolamine, *J. Biol. Chem.*, 256, 7687, 1981.

153. **Miller, M. A. and Kent, C.,** Characterization of the pathways for phosphatidylethanolamine biosynthesis in Chinese hamster ovary mutant and parental cell lines, *J. Biol. Chem.*, 261, 9753, 1986.

154. **Bjerve, K. S.,** The phospholipid substrates in the Ca^{2+}-stimulated incorporation of nitrogen bases into microsomal phospholipids, *Biochim. Biophys. Acta*, 306, 396, 1973.

155. **Voelker, D. R.,** Phosphatidylserine functions as the major precursor of phosphatidylethanolamine in BHK-21 cells, *Proc. Natl. Acad. Sci. U.S.A.*, 81, 2669, 1984.

156. **Baranska, J.,** A new pathway for phosphatidylserine synthesis in rat liver microsomes, *FEBS Lett.*, 228, 175, 1988.

157. **Kuge, O., Nishijima, M., and Akamatsu, Y.,** Isolation of a somatic cell mutant defective in phosphatidylserine biosynthesis, *Proc. Natl. Acad. Sci. U.S.A.*, 82, 1926, 1985.

158. **Kuge, O., Nishijima, M., and Akamatsu, Y.,** Phosphatidylserine biosynthesis in cultured Chinese hamster ovary cells. Isolation and characterization of phosphatidylserine auxotrops, *J. Biol. Chem.*, 261, 5790, 1986.

159. **Irving, H. R. and Exton, J. H.,** Phosphatidylcholine breakdown in rat liver plasma membranes, *J. Biol. Chem.*, 262, 3440, 1987.

160. **Lindmar, R., Löfelholz, K., and Sandman, J.,** On the mechanism of muscarinic hydrolysis of choline phospholipids in the heart, *Biochem. Pharmacol.*, 37, 4689, 1988.

161. **Prochazka, M., Hozak, U., and Kozak, L. D.,** A glycerol-3-phosphate dehydrogenase null mutant in BALB/cHcA mice, *J. Biol. Chem.*, 264, 4679, 1989.

162. **Sanghera, J. S. and Vance, D. E.,** CTP:phosphocholine cytidylyltransferase is a substrate for cAMP-dependent protein kinase *in vitro*, *J. Biol. Chem.*, 264, 1215, 1989.

163. **Agwu, D. E., McPhail, L. C., Chabot, M. C., Daniel, L. W., Wykle, R. L., and McCall, C. E.,** Choline-linked phosphoglycerides. A source of phosphatidic acid and diacylglycerol in stimulated neutrophils, *J. Biol. Chem.,* 264, 1405, 1989.

164. **Kiss, Z. and Anderson, W. B.,** Phorbol ester stimulates the hydrolysis of phosphatidylethanolamine in leukemic HL-60, NIH 3T3 and baby hamster kidney cells, *J. Biol. Chem.,* 264, 1483, 1989.

165. **Irona, H., Yoshioka, T., and Hotta, Y.,** Diacylglycerol kinase defect in a *Drosophila* retinal degeneration mutant rdgA, *J. Biol. Chem.,* 264, 5996, 1989.

166. **Weinhold, P. A., Rounsifer, M. E., and Feldman, D. A.,** The purification and characterization of CTP:phosphorylcholine cytidylyltransferase from rat liver, *J. Biol. Chem.,* 261, 5104, 1986.

167. **Feldman, D. A. and Weinhold, P. A.,** CTP:phosphorylcholine cytidylyltransferase from rat liver. Isolation and characterization of the catalytic subunit, *J. Biol. Chem.,* 262, 9075, 1987.

168. **Cornell, R.,** Chemical cross-linking reveals a dimeric structure for CTP:phosphocholine cytidylyltransferase, *J. Biol. Chem.,* 264, 9077, 1989.

169. **Pessin, M. S. and Raben, D. M.,** Molecular species analysis of 1,2-diglycerides stimulated by α-thrombin in cultured fibroblasts, *J. Biol. Chem.,* 264, 8729, 1989.

170. **Billah, M. M., Pai, J. K., Mullmann, T. J., Egan, R. W., and Siegel, M. I.,** Regulation of phospholipase D in HL-60 granulocytes. Activation by phorbol esters, diglyceride, and calcium ionophore via protein kinase C independent mechanisms, *J. Biol. Chem.,* 264, 9069, 1989.

171. **Macara, I. G.,** Elevated phosphocholine concentration in *ras*-transformed NIH 3T3 cells arises from increased choline kinase activity, not from phosphatidylcholine breakdown, *Mol. Cell. Biol.,* 9, 325, 1989.

172. **Hata, Y., Ogata, E., and Kojima, I.,** Platelet-derived growth factor stimulates the synthesis of 1,2-diacylglycerol from monoacylglycerol in BALB/C 3T3-cells, *Biochem. J.,* 262, 947, 1989.

Chapter 7

INOSITOL PHOSPHATIDES AND TRANSMEMBRANE SIGNALING

José M. Mato and Isabel Varela

TABLE OF CONTENTS

I. SYNTHESIS OF CDP-DIACYLGLYCEROL

Phosphatidic acid can be converted *in vivo* to the liponucleotide CDP-diacylglycerol in a reaction that requires CTP and magnesium (Figure 1).[1-4] Although dCTP has also been used *in vitro*, dCDP-diacylglycerol has not been isolated from liver extracts. The mammalian phosphatidic acid:CTP cytidylyltransferase has been detected in microsomes and mitochondria from several tissues, including liver, brain, and lung.[1-4] Whether the microsomal and mitochondrial enzymes contribute to the same or different pools of CDP-diacylglycerol is not known. The synthase of CDP-diacylglycerol has been reported to show little specificity with respect to the fatty acid composition of the phosphatidic acid used as substrate in the reaction,[5] which contrasts with the observation that phosphatidylinositol (a product of CDP-diacylglycerol metabolism) is rich in arachidonic acid (see Chapter 2). CDP-diacylglycerol is the common precursor for the synthesis of phosphatidylinositol, phosphatidylglycerol, and cardiolipin (Figure 1). A CDP-diacylglycerol hydrolase has been found in mammalian brain[6] and liver,[7] and might be an important factor in controlling the levels of CDP-diacylglycerol and, consequently, the formation of acidic phospholipids. Phosphatidylglycerol is formed in the mitochondria, where it is mainly localized, by the sequential action of phosphatidylglycerol phosphate synthase, which transfers a phosphatidyl unit from CDP-diacylglycerol to glycerol-3-phosphate, and a specific phosphatase.[8,9] With the exception of the adult pulmonary lung where phosphatidylglycerol is found in considerable amounts,[10] this phospholipid is only present as a trace intermediate in the synthesis of cardiolipins. Cardiolipin, a major component of the inner mitochondria, is also formed in this organelle by the condensation of CDP-diacylglycerol with phosphatidylglycerol.[9] The synthesis of phosphatidylinositol, which occurs by the action of CDP-diacylglycerol:inositol phosphatidyltransferase, was first thought to occur only in the endoplasmic reticulum, then to be transported to the plasma membrane.[11] More recently, however, phosphatidylinositol synthesis has been detected in both the endoplasmic reticulum and the plasma membrane of rat pituitary GH_3 cells.[12,13] Both synthase activities are inhibited by phosphatidylinositol, the product of the reaction. Since both degradation and resynthesis of phosphatidylinositol occur during stimulation of GH_3 cells with tyrotropin-releasing hormone,[12] it has been proposed that activation of phosphatidylinositol synthase takes place by release of the enzyme from inhibition by its product.[13]

II. SYNTHESIS OF PHOSPHATIDYLINOSITOL-4-PHOSPHATE AND PHOSPHATIDYLINOSITOL-4,5-BISPHOSPHATE

Plasma membrane phosphatidylinositol can be further phosphorylated to form phosphatidylinositol-4-phosphate and phosphatidylinositol-4,5-bisphosphate (Figure 2). More recently, evidence has been given in favor of the existence of a novel phosphoinositide containing four phosphates, phosphatidylinositol trisphosphate.[14] This novel phospholipid was detected in activated neutrophils, but not in resting cells.[14] Phosphatidylinositol-4-phosphate is formed by the action of a phosphatidylinositol kinase that uses ATP and phosphatidylinositol as substrates. Phosphatidylinositol kinase has been found associated to a variety of membrane structures, including lysosomes,[15] Golgi,[16] endoplasmic reticulum,[17] nuclear[18] and plasma[17] membranes, and coated vesicles.[19] Two types of phosphatidylinositol kinases, with a size of 55 and 230 kDa, respectively, have been separated from bovine brain by rate-zonal sucrose-gradient centrifugation.[20] Phosphatidylinositol kinase has also been partially purified from brain myelin[21] and murine livers.[22] The partially purified phosphatidylinositol kinase from murine livers does not catalyze the formation of phosphatidylinositol-4,5-bisphosphate.[22] A phosphatidylinositol kinase has been purified to near homogeneity from A431 cells.[23] The purified enzyme has an M_r of 55 kDa, as estimated by SDS gel

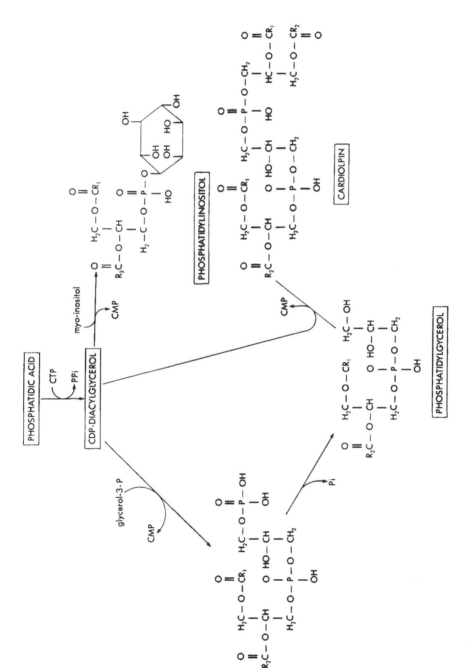

FIGURE 1. CDP-diacylglycerol is the common precursor for the synthesis of phosphatidylglycerol, cardiolipin, and phosphatidylinositol.

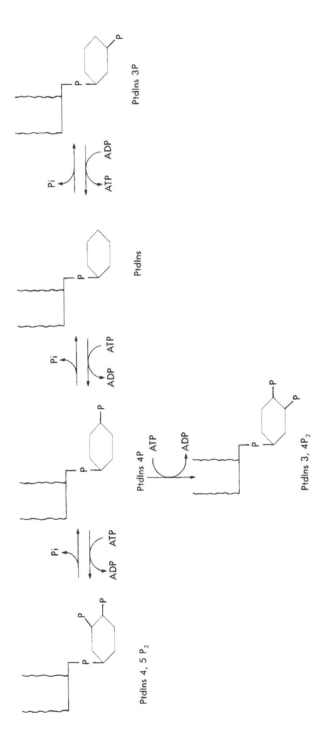

FIGURE 2.   Summary of routes by which inositol phospholipids are synthesized. PtdIns, phosphatidylinositol; PtdIns 4P, phosphatidylinositol-4-phosphate; PtdIns 3P, phosphatidylinositol-3-phosphate; PtdIns 4,5P₂, phosphatidylinositol-4,5-bisphosphate; PtdIns 3,4P₂, phosphatidylinositol-3,4-bisphosphate.

electrophoresis, appears to be active as a monomer, and is inhibited by micromolar concentrations of adenosine. This phosphatidylinositol kinase seems to be activated in response to EGF in A431 cells, although the mechanism by which EGF might activate this enzyme is unknown.[23] Membrane-bound phosphatidylinositol kinases with an M_r of about 80 and 53 kDa have been purified from rat brain[156] and human red blood cells,[174] respectively. Phosphatidylinositol kinase is known to associate with the PDGF receptor in 3T3 cells, and its association with the receptor increases 10- to 50-fold in PDGF-stimulated cells.[103] Analysis of cells bearing mutant PDGF receptors indicated that all mutants defective in PDGF-stimulated phosphatidylinositol kinase were also defective in PDGF-induced mitogenesis.[103] Mutation of the PDGF receptor by deletion of its kinase insert region resulted in the loss of PDGF-stimulated phosphatidylinositol kinase and PDGF-induced mitogenesis, but preserved the effect of PDGF on phosphatidylinositol hydrolysis.[103] These results indicate that the receptor-associated phosphatidylinositol kinase is not necessary to obtain phosphatidylinositol hydrolysis, but might be necessary to induce mitogenesis. Phosphatidylinositol-3-phosphate, but not the most common isomer phosphatidylinositol-4-phosphate, has been identified as a product of the phosphatidylinositol kinase associated with middle T-pp60[c-src] complexes from polyoma-infected fibroblasts[181,182] and with a number of cell surface receptors containing tyrosine kinase activity (Figure 2). Whether this lipid is involved in cellular growth and transformation remains to be determined. In addition to this phosphatidylinositol-3-kinase activity, anti-phosphotyrosine immunoprecipitates from PDGF-stimulated smooth muscle cells contain kinases that use as substrates phosphatidylinositol-4-phosphate and phosphatidylinositol-4,5-bisphosphate.[186] The products of these phosphorylations are phosphatidylinositol-3,4-bisphosphate and phosphatidylinositol trisphosphate (Figure 2).[186] Again, the physiological role of these lipids remains to be determined. In brain, phosphatidylinositol kinase is regulated by GTP analogs,[175] and in *Saccharomyces cerevisiae*, the results obtained with certain mutants with defective cAMP-cascade systems suggest that cAMP plays an important role in phosphoinositide synthesis through the phosphorylation of phosphatidylinositol and phosphatidylinositol-4-phosphate kinase.[176]

Elevated phosphatidylinositol kinase activity has been measured in rapidly growing cells, such as fibroblasts infected with Rous sarcoma virus[25] and rat mammary tumors,[26] and during hepatocarcinogenesis[24] and *Dictyostelium discoideum* differentiation.[27] As mentioned above, several protein-tyrosine kinases, including the viral oncogene product p60,[28,182] p68,[29] a *v-src*-related tyrosine kinase,[30] and the insulin-, PDGF-, and colony-stimulating factor 1 receptor tyrosine kinase,[31,103,181,186,206] have been reported to be associated with phosphatidylinositol-3-kinase activities. Present evidence indicates that both activities are due to separated enzymes,[32-35] and it has been proposed that these phosphatidylinositol kinase activities might be involved in cell growth and transformation. Formation of phosphatidylinositol-phosphate and phosphatidylinositol-bisphosphate by the addition of cAMP-dependent protein kinase has been reported in a variety of membrane preparations, including plasma membranes from lymphocytes,[36] and granulocytes,[37] and rabbit heart sarcoplasmic reticulum.[38] Phosphorylation of phospholamban, a putative regulator of Ca^{2+} transport by cardiac sarcoplasmic reticulum, by the catalytic subunit of cAMP-dependent protein kinase has been found to be associated with increased formation of phosphatidylinositol phosphate and phosphatidylinositol bisphosphate (the purified phospholamban contained a variety of phospholipids, including phosphatidylinositol).[39] Whether these effects are mediated directly by the catalytic subunit of the cAMP-dependent protein kinase or by endogenous phosphatidylinositol kinases is not known. Furthermore, it is also not known whether *in vivo* cAMP-dependent protein kinases regulate phosphatidylinositol kinase activity.

III. INOSITOL PHOSPHATIDES AND CELLULAR SIGNALING

In animal cells, phosphatidylinositol is present in all subcellular membranes and usually

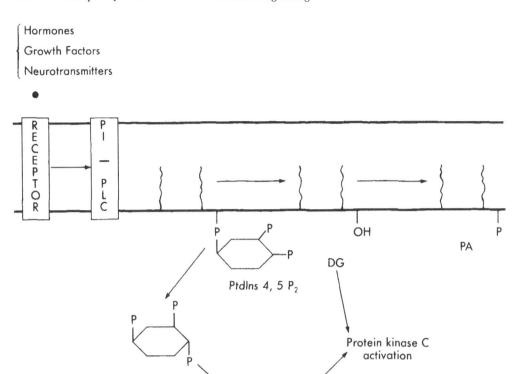

FIGURE 3.   Summary of the initial steps in the activation of phosphatidylinositol turnover in response to extra-cellular signals. PtdIns 4,5P$_2$, phosphatidylinositol-4,5-bisphosphate; DG, diacylglycerol; PA, phosphatidic acid; PI-PlC, phosphatidylinositol-specific phospholipase C.

accounts for less than 10% of the total cellular phospholipids (see Chapter 2). Most of the phosphatidylinositol, however, is in the endoplasmic reticulum and a small percent is at the plasma membrane. As previously mentioned, phosphatidylinositol shows the interesting property of being further phosphorylated to form the polyphosphoinositides phosphatidyli-nositol phosphate and phosphatidylinositol bisphosphate. The polyphosphoinositides are mainly localized at the plasma membrane and usually account for less than 10% of the total phosphatidylinositol pool.[40] Hokin and Hokin gave evidence for the first time of an increased turnover of the inositol phospholipids in response to acetylcholine or carbamylcholine.[41,42] Since then, the number of agents that have been shown to stimulate the turnover of the inositol lipids are legion, and the list includes hormones, neurotransmitters, growth factors, and oncogenes.[40] The pool of inositol lipids that is sensitive to hormones, growth factors, etc. seems to be limited to the plasma membrane.[43] The initial step in the activation of phosphatidylinositol turnover in response to extracellular signals seems to be the activation of a phosphatidylinositol-specific phospholipase C (PI-PlC) that hydrolyzes these phospho-lipids into diacylglycerol and inositol phosphates (Figure 3).[44,45] During cell stimulation, there is also an enhanced resynthesis of the inositol phosphatides to maintain appropriate levels of precursor lipids to guarantee the continued generation of diacylglycerol and inositol phosphates.[40] PI-PLC can cleave phosphatidylinositol, phosphatidylinositol-4-phosphate, and phosphatidylinositol-4,5-bisphosphate to yield a variety of inositol phosphates, including inositol-1-phosphate, inositol-1, 4-bisphosphate, inositol-1,4,5-trisphosphate, and inositol-1,3,4,5-tetrakisphosphate.[46,167]

A number of peaks with PI-PlC activity have been separated from a variety of tissue extracts.[47-50] Two immunologically different enzymes have been purified from seminal vesicular glands.[51,52] These enzymes have M_rs of 70 and 85 kDa, respectively, when measured by gel filtration.[51,52] A PI-PlC with an M_r of 143 kDa has been purified to homogeneity from bovine platelets.[53] One of the six different PI-PlC activities detected in bovine brain has also been purified to homogeneity and shows an M_r of 88 kDa.[54] Furthermore, three distinct PI-PlC have been purified from bovine brain with M_rs of 150, 145, and 85 kDa, respectively.[55,56] The purification of three types of PI-PlC from rat liver cytosol with M_rs of 140, 71, and 87 kDa, respectively,[57,157] has also been reported. Finally, one of the two peaks of PI-PlC activity that were resolved when guinea pig uterus cytosolic proteins were chromatographed on a DEAE-sepharose column has also been purified to homogeneity and shows an M_r of 62 kDa on sodium dodecyl sulfate polyacrylamide gels.[101] The fact that PI-PlCs are much larger than other phospholipases, like phospholipase A_2, whose size varies between 12 and 18 kDa, and the fact that they show large variations in size suggests that a number of domains of PI-PlC serve regulatory functions. In the brain, the localization of four different phospholipase C isozymes has been examined by *in situ* hybridization with specific oligonucleotide probes.[183] The observation that the various isozymes display different localizations raises the possibility that the different PI-PlCs are coupled to different classes of agonist receptors.

An attempt to name PI-PlC systematically by using Greek letters to designate the phospholipase enzymes with different primary structures, assigning the letters according to the chronological order of their purification and using Arabic numerals after the Greek letters to designate PI-PlCs derived by proteolysis or alternative splicing, has recently been published.[170] According to this nomenclature, mammalian tissues contain at least five immunologically distinct PI-PlC enzymes that appear to be separated gene products.[170]

The first enzyme was cloned using antibodies against a PI-PlC purified from bovine brain (PI-PlC II; M_r, 148 kDa),[58] and from the amino acid sequence of several tryptic peptide fragments.[59] The second enzyme was cloned using antibodies against a PI-PlC purified from guinea pig uterus (PI-PlC I).[60] The sequences of the two PI-PlCs show few similarities. Whereas two regions of the isozyme from rat brain have significant amino acid similarities with the products of various tyrosine kinase-related oncogenes (*yes, src, fgr, abl, fps, fes,* and *tck*),[58,59] the amino acid sequence of PI-PlC I exhibits a high degree of homology to thioredoxins, small protein cofactors in thiol-dependent redox reactions.[60] In the case of the PI-PlC from rat brain, the homologous regions were located in the area of the tyrosine kinase-related oncogenes that is not essential for catalytic activity, but which might be involved in the regulation of the kinase function.[58,59] The homology of PI-PlC I and thioredoxins has been interpreted as suggestive of a role for thiol-dependent reactions in the regulation of the PI-PlC function.[60] The predicted amino acid sequence of PI-PlC I contains several potential sites of phosphorylation in serine or threonine.[60] The cloning and sequence of PI-PlC I and PI-PlC III from bovine brain has also been reported.[127] Comparison of these sequences with that of PI-PlC II shows two short regions of homology present in all three enzymes, which might represent the catalytic domain. PI-PlC I and PI-PlC III do not contain the two regions present in PI-PlC II which are similar to the nonreceptor tyrosine kinases, suggesting different modes of modulation. Finally, the cloning and sequence of a PI-PlC-154 which encodes a protein with an M_r of about 150 kDa has also been reported.[128] Two regions of homology were observed between PI-PlC-154 and PI-PlC II from bovine brain. PI-PlC-154, however, does not contain the region in PI-PlC II with homology to the nonreceptor tyrosine kinases, suggesting that both enzymes are distinctly regulated.

The mechanisms by which agonists regulate PI-PlC activity are not clearly understood. Recently, evidence has been given supporting the view that EGF stimulation of the receptor tyrosine kinase activity might activate phosphatidylinositol-4,5-bisphosphate hydrolysis through tyrosine phosphorylation of either a soluble PI-PlC or a tightly associated protein.[60] In these experiments, a tenfold activation of PI-PlC was observed within 1 min of EGF addition by immunoprecipitation with an antiphosphotyrosine antibody of extracts from A-431 cells.[60] Using antibodies against PI-PlC I, PI-PlC II, and PI-PlC III, evidence has been obtained indicating that EGF receptor kinase directly and specifically phosphorylates PI-PlC II on tyrosine *in vivo*.[169,188,189] In A-431 cells, both bradykinin and exogenous ATP also stimulate the hydrolysis of phosphatidylinositol-4,5-bisphosphate. However, only EGF stimulated PI-PlC II phosphorylation,[169] indicating the presence of additional mechanisms of control of phospholipase activity. Tyrosine phosphorylation of PI-PlC II by EGF has also been observed *in vitro* using PI-PlC II purified from bovine brain and the purified EGF receptor.[184,189] Similar results have been observed following treatment with PDGF of 3T3 mouse fibroblasts.[189] PI-PlC I and PI-PlC III were not phosphorylated by the purified EGF receptor, and PI-PlC II was not phosphorylated by the insulin receptor cytoplasmic kinase domain.[184,189] This suggests specificity for a given phospholipase C within the family of protein tyrosine kinase receptors.[184] The enzymatic activity of PI-PlC II was not modified *in vitro* by tyrosine phosphorylation,[184,189] and it is therefore not possible to conclude how phosphorylation of the enzyme might be related to its activity. Microinjection of PI-PlC I or PI-PlC II into quiescent NIH 3T3 cells (about 10,000 molecules per cell) has been shown to produce a time- and dose-dependent induction of DNA synthesis, and the microinjected cells display a morphology similar to transformed cells, suggesting a role for these two phospholipases in cell growth.[178]

A variety of experiments point to a role for guanine nucleotide binding proteins (G) in the inositol phosphatide signaling system.[61,62] G proteins are $\alpha\beta\tau$ heterotrimers that share a common set of β and τ subunits and differ in their α subunit.[63] Although similar, the α subunits are distinct gene products and are coupled to specific agonist receptors.[63] The evidence that the activation of PI-PlC is coupled to specific receptors via G proteins is based on the capacity of nonhydrolyzable GTP analogs to stimulate PI-PlC in permeabilized cells and isolated membranes, and in the ability of pertussis toxin to inhibit agonist-stimulated PI-PlC activity.[64-75,173] Pertussis toxin is known to ADP-ribosylate G_i and G_o,[63] and positive results with this toxin have been interpreted as suggestive of a role of G_i or G_o in agonist-mediated PI-PlC stimulation. Cholera toxin has also been shown to inhibit receptor-mediated phosphatidylinositol-4,5-bisphosphate hydrolysis in several cell types, including T cells,[76] rat pancreatic acinar tissue,[77] and Flow 9000 cells.[78] Since cholera toxin ADP-ribosylates G_s protein,[63] these results are suggestive of a role for G_s in some receptor-mediated phosphatidylinositol-4,5-bisphosphate hydrolysis. Finally, in some systems, neither pertussis toxin nor cholera toxin inhibit receptor-mediated phosphatidylinositol-4,5-bisphosphate hydrolysis.[68,79-82] In these systems, another GTP-binding protein has been postulated to be involved,[63] although the possibility cannot be excluded that the interaction of the receptor with PI-PlC might not be mediated by G proteins. In *Drosophila*, the *norpA* (no receptor potential A) gene encodes a PI-PlC activity present in the eye which is an essential component of the *Drosophila* phototransduction signaling system.[114] Furthermore, there is strong evidence indicating that in invertebrates phototransduction involves a light-dependent activation of PI-PlC which is coupled to a G protein.[115,116] To investigate if a single cell possesses multiple G proteins that modulate PI-PlC activity but that are coupled to different receptors, CHO cells lacking muscarinic acetylcholine receptors have been transfected with individual muscarinic acetylcholine receptor subtypes and the pathway used by each of these receptors to activate phosphatidylinositol hydrolysis investigated.[185] The results indicate the existence of different pathways in a given cell that couple distinct G proteins to phosphatidylinositol hydrolysis.[185]

A number of studies have investigated the possible role of the GTP binding activity encoded by *ras* (proto)oncogenes (p21) in inositol phosphatide metabolism. p21*ras* binds and hydrolyzes GTP, as do G proteins.[83-85] p21*ras*, however, is a monomeric protein which does not interact with $\beta\tau$ subunits of other G proteins and there is no evidence in favor of the existence of *ras*-specific β and τ equivalent to those in G proteins.[86] p21*ras* does not seem to interact with adenylate cyclase in cells from higher organisms.[87,88] Although it has been reported that in various systems p21*ras* controls the activity of PI-PlC,[89-92] it has also been proposed that p21*ras* controls the activity of phospholipase A_2,[93,94] an enzyme that might be regulated by the $\beta\tau$ subunits of G protein,[95] and yet others have reported an inhibition of both PI-PlC and phospholipase A_2[96,97] or an activation of protein kinase C.[205] It has also been suggested that in *ras*-transformed cells, diacylglycerol is generated from a source other than inositol phosphatides.[98,205] More recently, evidence has been given indicating that elevated phosphocholine concentration in *ras*-transformed NIH 3T3 cells arises from increased choline kinase activity, and not as a consequence of increased phosphatidylinositol or phosphatidylcholine turnover.[190] In 3T3 cells, the activity of cellular *ras* has been neutralized by microinjection of a specific antibody.[99] Stimulation of these cells with phorbol ester and a Ca^{2+} ionophore, a treatment designed to imitate the action of phospholipase C or phospholipase A_2, did not prevent the inhibition of proliferation triggered by the antibody.[100] These data, therefore, do not support the idea that *ras* proteins control the action of phospholipases, and it has been proposed that phospholipid metabolism might be a biochemical link between growth factor receptors and cellular *ras* activity.[100] The occurrence of similar changes in the inositol phosphatide metabolism of cells transformed by *ras* and other cytoplasmic and membrane-associated oncogenes (*src*, *met*, *trk*, *mos*, and *raf*),[94] and the finding that an antibody to phosphatidylinositol-4,5-bisphosphate inhibits oncogene-induced mitogenesis,[158] further suggests the existence of common signaling pathways between cytoplasmic and membrane-associated oncogenes. Interestingly, it has been reported that several phospholipids involved in cellular signaling (i.e., phosphatidylinositol and phosphatidic acid) inhibited the stimulation of *ras* GTPase activity by a cytoplasmic-activating protein.[102] This effect was observed with phosphatidylinositol and phosphatidic acid species containing arachidonic acid, but not with those species containing long, saturated fatty acids or with other phospholipids, such as phosphatidylcholine, phosphatidylethanolamine, or diacylglycerol.[102] These results raise the possibility that the interaction of phospholipids with the GTPase-activating protein might be important in controlling *ras* activity during mitogenic stimulation.

IV. INOSITOL PHOSPHATES

Inositol phosphatides serve as precursors of two types of cellular messengers or mediators; inositol phosphates and diacylglycerol (Figure 2). An increased hydrolysis of phosphatidylinositol-4,5-bisphosphate to form inositol-1,4,5-trisphosphate, $Ins(1,4,5)P_3$, and diacylglycerol has been detected in many different cells in response to a large variety of signals, including hormones, growth factors, and neurotransmitters.[104] The increase in cellular $Ins(1,4,5)P_3$ has been shown to either precede or coincide with the starting of the Ca^{2+} signal,[105-108] which is consistent with the hypothesis that $Ins(1,4,5)P_3$ mediates Ca^{2+} mobilization. The time course of inositol phosphate generation varies among cells. Whereas in platelets the hydrolysis of inositol phosphatides in response to thrombin is completed within a few minutes of the addition of the signal,[109] in GH_3 pituitary cells stimulated with thyrotropin-releasing hormone the hydrolysis of phosphatidylinositol continues for about 30 min.[45] After being formed, $Ins(1,4,5)P_3$ is further metabolized to form additional inositol phosphates. $Ins(1,4,5)P_3$ can be dephosphorylated by a 5-phosphomonoesterase to form $Ins(1,4)P_2$. This enzyme has been detected in the particulate and soluble fractions of different cells.[110,111] The enzyme isolated from platelets is phosphorylated and activated severalfold by protein kinase C.[112,113] $Ins(1,4,5)P_3$ can also be phosphorylated to yield $Ins(1,3,4,5)P_4$, by a kinase which might require calcium ions and calmodulin, then to be dephosphorylated to form $Ins(1,3,4)P_3$, $Ins(3,4)P_2$, and $Ins(1,3)P_2$.[117,118] The inositol tetrakisphosphates

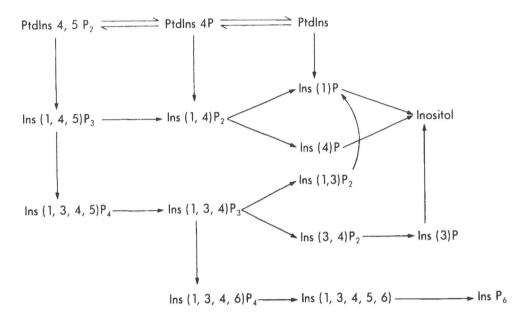

FIGURE 4.   Current knowledge of the metabolism of inositol phosphates derived from inositol-1,4,5-trisphosphate [Ins(1,4,5)P_3].

Ins(1,3,4,6)P_4 and Ins(3,4,5,6)P_4 have been identified in bovine adrenal glomerulosa cells, the levels of these molecules being rapidly increased by angiotensin II.[179,204] The levels of an inositol tetrakisphosphate, probably Ins(3,4,5,6)P_4, are markedly increased in fibroblasts transformed with the *v-src* oncogene, whereas the levels of inositol trisphosphate and other inositol tetrakisphosphates were not modified.[191] Inositol pentakis- and hexakisphosphates are not confined to plants and avian erythrocytes and have been found in many different systems,[119-121,179] and in pituitary cells stimulated with gonadotropin-releasing hormone, an increase in the cellular content of both inositol pentakis- and hexakisphosphate has been observed.[122] Ins(1,3,4,5,6)P_5 regulates the affinity of hemoglobin for oxygen in avian erythrocytes, and introduction of InsP$_6$ into viable human or porcine erythrocytes improved tissue oxygenation when normal blood flow was impaired.[129] Infusion of inositol pentakis- and hexakisphosphate into a discrete brain stem nucleus implicated in cardiovascular regulation has been shown to result in changes in heart rate and blood pressure.[121] The pathways for the synthesis of inositol pentakis- and hexakisphosphate has been shown to be independent of the other inositol phosphates discussed in this chapter.[193,194] Ins(1,4)P_2 can be converted to Ins 4-P by a phosphomonoesterase that is inhibited by lithium ions, which can explain why Ins(1,4)$_2$ and Ins(1,3,4)P_3 accumulate in cells treated with lithium ions.[104] In addition, Ins 3-P, Ins 4-P, and inositol-1,2-cyclic monophosphate are also found in cells as a result of inositol phosphate metabolism.[46] Although the details of the pathways used to synthesize the various inositol phosphates are not known in detail, Figure 4 summarizes our current knowledge of the metabolism of inositol phosphates derived from Ins(1,4,5)P_3.

The hydrolysis of inositol phosphatides occurs mainly in response to calcium-mobilizing agonists that have an effect on both internal and external calcium pools. The initial response of the cell to these agents is a release of internal Ca^{2+}, which is followed by the entry of Ca^{2+} through the cell plasma membrane. Ins(1,4,5)P_3 have been shown to release Ca^{2+} from intracellular stores, presumably from the endoplasmic reticulum, in many nonmuscle cells, including pancreatic cells, hepatocytes, neutrophils, GH$_3$ cells, Swiss 3T3 cells, and platelets.[104] The first evidence was obtained using permeabilized pancreatic cells.[123] In this system, the release of Ca^{2+} was monitored with a calcium electrode following the addition of Ins(1,4,5)P$_3$. A variety of signals, including many hormones, growth factors, and neurotransmitters, induce in their target cells oscillations in the internal free calcium which are mediated by Ins(1,4,5)P$_3$.[171] Interestingly, these pulses of intracellular Ca^{2+} occur even when the concentration of Ins(1,4,5)P$_3$ is held constant, either by the continuous application through a patch pipette of Ins(1,4,5)P$_3$ or its nonmetabolizable analog inositol trisphosphorothioate.[172] This suggests that pulsatile intracellular Ca^{2+} release is due to intermittent modulation of Ca^{2+} channels of intracellular reservoirs which are obtained at constant concentrations of Ins(1,4,5)P$_3$.[172] In muscle cells, Ins(1,4,5)P$_3$ activates a calcium channel from vascular smooth muscle sarcoplasmic reticulum.[124] In addition, Ins(1,4,5)P$_3$ has also been shown to open dihydropyridine calcium channels of transverse tubules from skeletal muscle plasma membrane,[125] which suggests a major role for this inositol trisphosphate in the modulation of muscle contraction. In smooth muscle, the rate of force development after photolytic release of Ins(1,4,5)P$_3$ is also compatible with a role for this inositol trisphosphate in pharmacomechanical coupling.[192] Evidence has been given suggesting a role for Ins(1,3,4,5)P$_4$ in modulating the entry of Ca^{2+} across the plasma membrane, using eggs of the sea urchin *Lytechninus variegatus*.[126] Ins(1,3,4,5)P$_4$ was only effective if Ins(2,4,5)P$_3$ was co-injected and if Ca^{2+} was present in the incubation media. This synergism of the two inositol phosphates is not a general phenomenon, and in eggs from *L. pictus* and *Psammechinus milaris*, activation occurs by microinjection of Ins(2,4,5)P$_3$, without synergism, with Ins(1,3,4,5)P$_4$.[195,196] A synergism between inositol trisphosphate and Ins(1,3,4,5)P$_4$ has been reported in cells from mouse lacrimal glands,[197] and on the release of Ca^{2+} from mouse pituitary microsomes.[198] Moreover, Ins(1,3,4,5)P$_4$ has been shown to inhibit the re-uptake of Ca^{2+} into internal stores and to cause the inhibition of Ins(1,4,5)P$_3$-5-phosphatase.[166] The hypothesis is that Ins(1,4,5)P$_3$ would release Ca^{2+} from intracellular stores, then to be converted to Ins(1,3,4,5)P$_4$, which would control Ca^{2+} passage between different cellular compartments or the entry of Ca^{2+} through the plasma membrane.[199] Ins(1,4,5)P$_3$ is thought to act at receptor sites on the endoplasmic reticulum to trigger Ca^{2+} release. Ins(1,4,5)P$_3$ receptors have been identified in peripheral tissues and brain,[159-162] and an Ins(1,4,5)P$_3$ binding protein with an M$_r$ of 260 kDa has been purified to homogeneity.[162] Phosphorylation of this binding protein with cAMP-dependent protein kinase reduced the ability of Ins(1,4,5)P$_3$ to release Ca^{2+}.[163] The properties of the rat liver Ins(1,4,5)P$_3$ receptor have been examined in response to various hormones.[180] Present evidence suggests the existence of two states of the Ins(1,4,5)P$_3$ receptor, one with high affinity and a second with low affinity. The high/low affinity ratio is increased by hormones that stimulate the generation of Ins(1,4,5)P$_3$, and this effect is reversed by hormones that act via cAMP.[180] These results support the view of the existence of cross-talk between these two second messengers. Purified Ins(1,4,5)P$_3$ receptor has recently been reconstituted into lipid vesicles.[201] Ins(1,4,5)P$_3$ and other inositol phosphates have been shown to stimulate calcium flux in the reconstituted vesicles with potencies and specificities that parallel the calcium-releasing actions of Ins(1,4,5)P$_3$. These results strongly indicate that the purified receptor mediates both the recognition of Ins(1,4,5)P$_3$ and calcium transport. The cloning and sequence of the inositol (1,4,5)P$_3$ receptor from cerebellar Purkinje neurons has also been reported recently.[202] Surprisingly, the sequence of the Ins(1,4,5)P$_3$ receptor is very similar to the recently cloned ryanodine receptor (RR),[202,203] the Ca^{2+} channel of the sarcoplasmic reticulum of skeletal muscle.

V. REGULATION OF PROTEIN KINASE C BY DIACYLGLYCEROL

Protein kinase C occurs in two forms, as a native 80-kDa protein and as a smaller proteolytic fragment.[130] Whereas the native protein kinase C is activated by the simultaneous presence of phospholipid, diacylglycerol, and Ca^{2+}, the proteolytic fragment phosphorylates protein substrates in the absence of these regulators, indicating that a modulatory fragment has been lost.[130] The phospholipid-protein kinase C interaction does not seem to be highly specific.[131] Among the different phospholipids of biological interest, there seems to be an absolute requirement for phosphatidylserine, although the nature of the acyl groups of the phosphatidylserine does not play an important role.[130] Phosphatidylethanolamine or phosphatidylcholine alone do not regulate protein kinase C, but both can form active mixtures with phosphatidylserine. N-methylation of phosphatidylethanolamine decreases the ability of a mixture of phosphatidylethanolamine and phosphatidylserine to activate protein kinase C.[132] In the presence of diacylglycerol, the affinity of protein kinase C for Ca^{2+} increases, so that it can be activated at physiological basal levels of this cation.[130] The interaction diacylglycerol-protein kinase C is stoichiometric and only the *sn*-1,2-diacylglycerols activate protein kinase C, whereas the *sn*-2,3-diacylglycerol enantiomer and the 1,3-diacylglycerol diastereomer are inactive.[132] There is no marked specificity about the nature of the acyl groups of the diacylglycerol molecule. Thus, 1-oleoyl-2-acetyl glycerol and 1-acetyl-2-oleoyl glycerol are both potent activators of protein kinase C.[133] Diacylglycerols with a saturated fatty acid at position 1 and an unsaturated fatty acid at position 2 are thought to be the natural modulators of protein kinase C.[130] It seems, therefore, that as long as the alkyl chains of the diacylglycerol provide a hydrophobic environment, they can sustain the activation of protein kinase C. Modifications at the glycerol backbone of the diacylglycerol molecule markedly inhibit the protein kinase C-activating activity. Thus ether-for-acyl substitutions have been shown to decrease the activity of the resulting diacylglycerol,[134,135] which might be of physiological importance in light of the relatively large proportion of cellular phospholipids having a 1-alkyl structure.

Molecular cloning analysis has shown that protein kinase C is a family of multiple subspecies having closely related structures. At least four different functional subspecies, three closely related subspecies, and one distantly related species have been identified.[136-144,173] Whether all these members of the protein kinase C family have different functions, regulate, or act on different substrates is not known. Interestingly, different cell types express restricted patterns of mRNAs, indicating that expression of the different subspecies is regulated at the level of transcription.[136-140,173] In conclusion, the present evidence indicates that elevation of cellular diacylglycerol as a result of the hydrolysis of a variety of phospholipids (i.e., phosphatidylcholine and phosphatidylinositol) activates a family of protein kinases C which might act on specific substrates, depending on the signal and on the cellular system. Finally, it is interesting to note the similarity between a 100-residue segment in the carboxy-terminal half of bovine brain PI-PlC and the regulatory domain of rat protein kinase C-I.[168] The biological function of these two regions, however, is not known.

Protein kinase C exists in both cytosolic and plasma membrane-bound forms, the cytosolic form of the enzyme being the most abundant under resting conditions.[130] However, upon cell stimulation and elevation of cellular Ca^{2+}, there is a net translocation of the enzyme to the plasma membrane. Activation of protein kinase C can take place by elevation of Ca^{2+} levels and by diacylglycerol (Figure 3).[130] Protein kinase C phosphorylates proteins in serine and threonine residues, and has been shown to phosphorylate and modulate the activity of a large number of enzymes, including, glycogen phosphorylase kinase, glycogen kinase, myosin light-chain kinase, cytochrome P450, phospholipid methyltransferase, and phospholipase C.[130,150] This broad protein substrate specificity suggests that this enzyme is involved in many cellular processes. Protein kinase C seems to play an important role in cell growth. Tumor-promoting phorbol esters, which can translocate to the plasma membrane and activate protein kinase C in a fashion similar to diacylglycerol,[146] are known to stimulate cell growth in a variety of systems.[130] Moreover, overexpression of transfected protein kinase C genes in 3T3 cells and rat fibroblasts results in abnormal cellular growth.[147] Down-regulation of protein kinase C, by long-term incubation with phorbol esters, abolishes cell proliferation in response to a new addition of phorbol esters, to be restored by microinjection of purified protein kinase C.[148] Furthermore, protein kinase C phosphorylates a number of receptors implicated in the control of cell growth and differentiation as well as in the modulation of the immune response.[130,150] In the nematode *Caenorhabditis elegans*, it has been observed that mutations in the *tpa*-1 gene, coding for a protein similar to protein kinase C in other animals, correct specific alterations in the behavior and development induced by tumor-promoting phorbol esters.[164] Since malignant transformation by *ras* and other onco-genes produces common alterations in inositol phospholipid signaling pathways[94] and an accumulation of diacylglycerol,[98] and a revertant of a *ras*-transformed NIH 3T3 cell ex-pressing high levels of p21*ras* protein had a reduced total protein kinase C activity and a normal distribution of the enzyme between the cytosol and membranes,[149] it is possible that protein kinase C might also play a role in the transformation process. Related to this hypothesis, the α-protein kinase C complementary DNA from a murine ultraviolet-induced fibrosarcoma cell line has been cloned and sequenced and found to have four-point mutations, three of them in the highly conserved regulatory domain and one in the conserved region of the catalytic domain.[207] Expression of this mutant α-protein kinase C gene in normal Balb/c 3T3 fibroblasts results in a fibrosarcoma-like protein kinase C localization, with 87% of the kinase activity associated with the membrane, and in cell transformation, suggesting that point mutations in the primary structure of protein kinase C modulate enzyme function and are responsible for inducing oncogenicity.[207] Finally, protein kinase C plays an important role in modulating the release of serotonin from platelets,[130] the exocytosis of cellular constituents in a variety of cells,[130] the generation of superoxide by neutrophils,[151] neuronal excitability,[152] and arachidonic acid release and prostaglandin synthesis.[153] GTP-binding proteins are also targets of protein kinase C, and phosphorylation might regulate G-protein function.[200]

In addition to activating protein kinase C, diacylglycerol is also a source of arachidonic acid, as the result of the action of a specific lipase. Furthermore, diacylglycerol functions as a chemoattractant in leukocytes[154] and an activator of glucose transport in adipocytes,[155] and reduces Ca^{2+} currents in chick sensory neurons,[165] when applied outside the cell. Interestingly, these extracellular actions of diacylglycerol seem to be independent of protein kinase C activation.[155,165] Diacylglycerol at physiological levels has been reported to increase the susceptibility of phospholipid membranes to fusion.[187] Moreover, the effect of phorbol esters on Ca^{2+} currents in chick sensory neurons might also be extracellular.[165] These results indicate that in the classical experiment in which phorbol esters or a diacylglycerol analog are applied outside the cell to test for their effects on a given cellular activity, a variety of control experiments should be carried out before concluding that a protein kinase C is involved.

REFERENCES

1. **Petzold, G. L. and Agranoff, B. W.**, Biosynthesis of cytidine diphosphate diglyceride by embryonic chick brain, *J. Biol. Chem.*, 242, 1187, 1967.
2. **Van Golde, L. M. G., Raben, J., Batenburg, J. J., Fleischer, B., Zambrano, F., and Fleischer, S.**, Biosynthesis of lipids in Golgi complex and other subcellular fractions from rat liver, *Biochim. Biophys. Acta*, 360, 179, 1974.
3. **Hallman, M. and Gluck, L.**, Phosphatidylglycerol in lung surfactant. II. Subcellular distribution and mechanism of biosynthesis *in vitro*, *Biochim. Biophys. Acta*, 409, 172, 1975.
4. **Harding, P. G. R., Chan, F., Casola, P. G., Fellows, G. F., Wong, T., and Possmayer, F.**, Subcellular distribution of the enzymes related to phospholipid synthesis in developing rat lung, *Biochim. Biophys. Acta*, 750, 373, 1983.

5. **Holub, B. J. and Pierkarski, J.,** Biosynthesis of molecular species of CDP-diglyceride from endogenously labelled phosphatide in rat liver microsomes, *Lipids*, 11, 251, 1976.

6. **Rittenhouse, H. G., Seguin, E. B., Fisher, S. K., and Agranoff, B. W.,** Properties of CDP-diglyceride hydrolase from guinea pig brain, *J. Neurochem.*, 36, 991, 1981.

7. **Nicholson, D. W. and McMurray, W. C.,** CDP-diglyceride hydrolase from pig liver mitochondria, *Can. J. Biochem. Cell. Biol.*, 62, 1205, 1984.

8. **Kiyasu, J. Y., Peringer, R. A., and Kennedy, E. P.,** The biosynthesis of phosphatidylglycerol, *J. Biol. Chem.*, 238, 2293, 1963.

9. **Hostetler, K. Y.,** Polyglycerophospholipids: phosphatidylglycerol, diphosphatidylglycerol and *bis*(monoacylglycerol)phosphate, in *Phospholipids*, Hawthorne, J. N. and Ansell, G. B., Eds., Elsevier, Amsterdam, 1982, 215.

10. **Possmayer, F.,** Pulmonary phosphatidate phosphohydrolase and its relation to the surfactant system of the lung, in *Phosphatidate Phosphohydrolase*, Brindley, D. N., Ed., CRC Press, Boca Raton, FL, 1987.

11. **Bell, R. M. and Coleman, R. A.,** Enzymes of glycolipid synthesis in eukaryotes, *Annu. Rev. Biochem.*, 49, 459, 1980.

12. **Imai, A. and Gershengorn, M. C.,** Independent phosphatidylinositol synthesis in pituitary plasma membrane and endoplasmic reticulum, *Nature*, 325, 726, 1987.

13. **Imai, A. and Gershengorn, M. C.,** Regulation by phosphatidylinositol of rat pituitary plasma membrane and endoplasmic reticulum phosphatidylinositol synthase activities, *J. Biol. Chem.*, 262, 6457, 1987.

14. **Traynor-Kaplan, A. E., Harris, A. L., Thompson, B. L., and Taylor, P.,** An inositol tetrakisphosphate-containing phospholipid in activated neutrophils, *Nature*, 334, 353, 1988.

15. **Collins, C. A. and Wells, W. W.,** Identification of phosphatidylinositol kinase in rat liver lysosomal membranes, *J. Biol. Chem.*, 258, 2130, 1983.

16. **Jergil, G. and Sundler R.,** Phosphorylation of phosphatidylinositol in rat liver Golgi, *J. Biol. Chem.*, 258, 7968, 1983.

17. **Cockcroft, S., Taylor, J. A., and Judah, J. D.,** Subcellular localisation of inositol lipid kinases in rat liver, *Biochim. Biophys. Acta*, 845, 163, 1985.

18. **Smith, C. D. and Wells, W. W.,** Phosphorylation of rat liver nuclear envelopes, *J. Biol. Chem.*, 258, 9368, 1983.

19. **Campbell, C. R., Fishman, J. B., and Fine, R. E.,** Coated vesicles contain phosphatidylinositol kinase, *J. Biol. Chem.*, 260, 10948, 1985.

20. **Endemann, G., Dunn, S. N., and Cantley, L. C.,** Bovine brain contains two types of phosphatidylinositol kinase, *Biochemistry*, 26, 6845, 1987.

21. **Saltiel, A. R., Fox, J. A., Sherline, P., Sahyoun, N., and Cuatrecasas, P.,** Purification of phosphatidylinositol kinase from bovine brain myelin, *Biochem. J.*, 241, 759, 1987.

22. **Suarez-Quian, C. A., O'Shea, J. J., and Klausner, R. D.,** Partially purified phosphatidylinositol kinase does not catalyze the formation of phosphatidylinositol-4,5-bisphosphate, *FEBS Lett.*, 143, 512, 1987.

23. **Walker, D. H., Dougherty, N., and Pike, L. J.,** Purification and characterization of a phosphatidylinositol kinase from A431 cells, *Biochemistry*, 27, 6504, 1988.

24. **Olson, J. W.,** Enhanced phosphatidylinositol kinase activity is associated with early stages of hepatocarcinogenesis and hepatocellular carcinoma, *Biochem. Biophys. Res. Commun.*, 132, 969, 1985.

25. **Sugimoto, Y. and Erikson, R.,** Phosphatidylinositol kinase activities in normal and rous sarcoma virus-transformed cells, *Mol. Cell. Biol.*, 5, 3194, 1985.

26. **Sharoni, Y., Teuerstein, I., and Levy, J.,** Phosphoinositide phosphorylation precedes growth in rat mammary tumors, *Biochem. Biophys. Res. Commun.*, 134, 876, 1986.

27. **Varela, I., Van Lookeren-Campagne, M. M., Alvarez, J. F., and Mato, J. M.,** The developmental regulation of phosphatidylinositol kinase in *Dictyostelium discoideum*, *FEBS Lett.*, 211, 64, 1987.

28. **Sugimoto, Y., Whitman, M., Cantley, L. C., and Erikson, R. L.,** Evidence that the rous sarcoma virus transforming gene product phosphorylates phosphatidylinositol and diacylglycerol, *Proc. Natl. Acad. Sci. U.S.A.*, 81, 2117, 1984.

29. **Macara, I. G., Marinetti, G. V., and Balduzzi, P. C.,** Transforming protein of avian sarcoma virus VR2 is associated with phosphatidylinositol kinase activity: possible role in tumorigenesis, *Proc. Natl. Acad. Sci. U.S.A.*, 81, 2728, 1984.

30. **Fukami, Y., Owada, M. K., Sumi, M., and Hayashi, F.,** A p60$^{v\text{-}src}$-related tyrosine kinase in the acetylcholine receptor-rich membranes of *Narke japonica*: association and dissociation of phosphatidylinositol kinase activity, *Biochem. Biophys. Res. Commun.*, 139, 473, 1986.

31. **Ruderman, N. B., Kapeller, R., White, M. R., and Cantley, L. C.,** Activation of phosphatidylinositol 3-kinase by insulin, *Proc. Natl. Acad. Sci. U.S.A.*, in press.

32. **Thompson, D. M., Cochet, C., Chambaz, E. M., and Gill, G. N.,** Separation and characterization of phosphatidylinositol kinase activity that co-purifies with the epidermal growth factor receptor, *J. Biol. Chem.*, 260, 8824, 1985.

33. **MacDonald, M. L., Kuenzel, E. A., Glomset, J. A., and Krebs, E. G.,** Evidence from two transformed cell lines that the phosphorylation of peptide tyrosine and phosphatidylinositol are catalyzed by different proteins, *Proc. Natl. Acad. Sci. U.S.A.,* 82, 3993, 1985.

34. **Fischer, S., Fagard, R., Comoglio, P., and Gacon, G.,** Phosphoinositides are not phosphorylated by the very active tyrosine kinase protein kinase from the murine lymphoma LSTRA, *Biochem. Biophys. Res. Commun.,* 132, 481, 1985.

35. **Carrascosa, J. M., Schleicher, E., Maier, R., Hackenberg, C., and Wieland, O. H.,** Separation of the protein-tyrosine kinase and phosphatidylinositol kinase activities of the human placental insulin receptor, *Biochim. Biophys. Acta,* 971, 170, 1988.

36. **Sarkadi, B., Enyedi, A., Farago, A., Meszaros, G., Kremmer, T., and Gardos, G.,** Cyclic AMP-dependent protein kinase stimulates the formation of polyphosphoinositides in lymphocyte plasma membrane, *FEBS Lett.,* 152, 195, 1983.

37. **Farkas, G., Enyedi, A., Sarkadi, B., Gardos, G., Nagy, Z., and Farago, A.,** Cyclic AMP-dependent protein kinase stimulates the phosphorylation of phosphatidylinositol to phosphatidylinositol 4-monophosphate in a plasma membrane preparation from pig granulocytes, *Biochem. Biophys. Res. Commun.,* 124, 871, 1984.

38. **Enyedi, A., Farago, A., Sarkadi, B., and Gardos, G.,** Cyclic AMP-dependent protein kinase and Ca^{2+}-calmodulin stimulate the formation of polyphosphoinositides in a sarcoplasmic reticulum preparation of rabbit heart, *FEBS Lett.,* 176, 235, 1984.

39. **Jakab, G. and Kranias, E. G.,** Phosphorylation and dephosphorylation of purified phospholamban and associated phosphatidylinositides, *Biochemistry,* 27, 3799, 1988.

40. **Downes, P. C. and Michell, R. H.,** Inositol phospholipid breakdown as a receptor-controlled generator of second messengers, in *Molecular Mechanisms of Transmembrane Signalling,* Cohen, P. and Houslay, M. D., Eds., Elsevier, Amsterdam, 1985.

41. **Hokin, M. R. and Hokin, L. E.,** Effects of acetylcholine on the turnover of phosphoryl units in individual phospholipids of pancreas slices and brain cortex, *Biochim. Biophys. Acta,* 18, 102, 1955.

42. **Hokin, M. R. and Hokin, L. E.,** The presence of phosphatidic acid in animal tissues, *J. Biol. Chem.,* 233, 800, 1958.

43. **Rana, R. S., Kowluru, A., and MacDonald, M. J.,** Secretagogue-responsive and -unresponsive pools of phosphatidylinositol in pancreatic islets, *Arch. Biochem. Biophys.,* 245, 411, 1986.

44. **Majerus, P. W., Connolly, T. M., Deckmyn, H., Ross, T. S., Bross, T. E., Ishii, H., Bansal, V. S., and Wilson, D. B.,** The metabolism of phosphoinositide-derived messenger molecules, *Science,* 234, 1519, 1986.

45. **Imai, A. and Gershengorn, M. C.,** Phosphatidylinositol 4,5-bisphosphate turnover is transient while phosphatidylinositol turnover is persistent in thyrotropin-releasing hormone-stimulated rat pituitary cells, *Proc. Natl. Acad. Sci. U.S.A.,* 83, 8540, 1986.

46. **Majerus, P. W., Connolly, T. M., Bansal, V. S., Inhorn, R. C., Ross, T. S., and Lips, D. L.,** Inositol phosphates: synthesis and degradation, *J. Biol. Chem.,* 263, 3051, 1988.

47. **Hirasawa, K., Irvine, R. F., and Dawson, R. M. C.,** Heterogeneity of the calcium-dependent phosphatidylinositol phosphodiesterase in rat brain, *Biochem. J.,* 205, 437, 1982.

48. **Low, M. G., Carroll, R. C., and Weglicki, W. B.,** Multiple forms of phosphoinositide-specific phospholipase C of different relative molecular masses in animal tissues, *Biochem. J.,* 221, 813, 1984.

49. **Nakanishi, H., Nomura, H., Kikkawa, U., Kishimoto, A., and Nishizuka, Y.,** Rat brain and liver soluble phospholipase C: resolution of two forms with different requirements for calcium, *Biochem. Biophys. Res. Commun.,* 132, 582, 1985.

50. **Banno, Y., Nakashima, S., and Nozawa, Y.,** Partial purification of phosphoinositide phospholipase C from human platelet cytosol: characterization of its three forms, *Biochem. Biophys. Res. Commun.,* 136, 713, 1986.

51. **Hofmann, S. L. and Majerus, P. W.,** Identification and properties of two distinct phosphatidylinositol-specific phospholipase C enzymes from sheep seminal vesicular glands, *J. Biol. Chem.,* 257, 6461, 1982.

52. **Wilson, D. B., Bross, T. E., Hofmann, S. L., and Majerus, P. W.,** Hydrolysis of polyphosphoinositides by purified sheep seminal vesicle phospholipase C enzymes, *J. Biol. Chem.,* 259, 11718, 1984.

53. **Hakata, H., Kambayashi, J., and Kosaki, G.,** Purification and characterization of phosphatidylinositol-specific phospholipase C from bovine platelets, *J. Biochem.,* 92, 929, 1982.

54. **Rebecchi, M. J. and Rosen, O. M.,** Purification of a phosphoinositide-specific phospholipase C from bovine brain, *J. Biol. Chem.,* 262, 12526, 1987.

55. **Ryu, S. H., Cho, K. S., Lee, K. Y., Suh, P. G., and Rhee, S. G.,** Purification and characterization of two immunologically distinct phosphoinositide-specific phospholipase C from bovine brain, *J. Biol. Chem.,* 262, 12511, 1987.

56. **Ryu, S. H., Cho, K. S., Lee, K. Y., Suh, P. G., and Rhee, S. G.,** Bovine brain cytosol contains three immunologically distinct forms of inositol phospholipid-specific phospholipase C, *Proc. Natl. Acad. Sci. U.S.A.,* 84, 6649, 1987.

57. **Nakanishi, O., Homma, Y., Kawasaki, H., Emori, Y., Suzuki, K., and Takenawa, T.,** Purification of two distinct types of phosphoinositide-specific phospholipase C from rat liver, *Biochem. J.,* 256, 453, 1988.

58. **Suh, P. G., Ryu, S. H., Moon, K. H., Suh, H. W., and Rhee, S. G.,** Inositol phospholipid-specific phospholipase C: complete cDNA and protein sequences and sequence homology to tyrosine kinase-related oncogene products, *Proc. Natl. Acad. Sci. U.S.A.,* 85, 5419, 1988.

59. **Stahl, M. L., Ferenz, C. R., Kelleher, K. L., Kriz, R. W., and Knopf, J. L.,** Sequence similarity of phospholipase C with the non-catalytic region of src, *Nature,* 332, 269, 1988.

60. **Bennett, C. F., Balcarek, J. M., Varrichio, A., and Crooke, S. T.,** Molecular cloning and complete amino-acid sequence of form-I phosphoinositide-specific phospholipase C, *Nature,* 334, 268, 1988.

60a. **Wahl, M. I., Daniel, T. O., and Carpenter, G.,** Antiphosphotyrosine recovery of phospholipase C activity after EGF treatment of A-431 cells, *Science,* 241, 968, 1988.

61. **Cockcroft, S.,** Phosphoinositide phosphodiesterase: regulation by a novel guanine nucleotide binding protein, Gp., *Trends Biochem. Sci.,* 12, 75, 1987.

62. **Lo, W. W. Y. and Hughes, J.,** Receptor-phosphoinositidase C coupling. Multiple G-proteins?, *FEBS Lett.,* 224, 1, 1987.

63. **Casy, P. J. and Gilman, A. G.,** G protein involvement in receptor-effector coupling, *J. Biol. Chem.,* 263, 2577, 1988.

64. **Katada, T. and Ui, M.,** Direct modification of the membrane adenylate cyclase system islet-activating protein due to ADP-ribosylation of a membrane protein, *Proc. Natl. Acad. Sci. U.S.A.,* 79, 3129, 1982.

65. **Smith, C. D., Lane, B. C., Kusaka, I., Verghese, M. W., and Snyderman, R.,** Chemoattractant receptor induced hydrolysis of phosphoinositide 4,5 bisphosphate in human polymorphonuclear leukocyte membranes, *J. Biol. Chem.,* 260, 5875, 1985.

66. **Cockcroft, S.,** Phosphoinositide phosphodiesterase: regulation by a novel guanine nucleotide binding protein, Gp., *Trends Pharmacol. Sci.,* 12, 75, 1987.

67. **Gonzales, R. A. and Crews, F. T.,** Guanine nucleotide stimulates production of inositol trisphosphate in rat cortical membranes, *Biochem. J.,* 232, 799, 1985.

68. **Straub, R. E. and Gershengorn, M. C.,** Thyrotropin-releasing hormone and GTP activate inositol trisphosphate formation in membranes isolated from rat pituitary cells, *J. Biol. Chem.,* 261, 2712, 1986.

69. **Holian, A.,** Leukotriene B$_4$ stimulation of phosphatidylinositol turnover in macrophages and inhibition by pertussis toxin, *FEBS Lett.,* 201, 15, 1986.

70. **Taylor, C. W., Merritt, J. E., Putney, J. W., Jr., and Rubin, R. P.,** A guanine nucleotide-dependent regulatory protein couples substance P receptors to phospholipase C in rat parotid gland, *Biochem. Biophys. Res. Commun.,* 136, 362, 1986.

71. **Baldassare, J. J. and Fisher, G. J.,** Regulation of membrane-associated and cytosolic phospholipase C activates in human platelets by GTP, *J. Biol. Chem.,* 261, 11942, 1986.

72. **Baldassare, J. J. and Fisher, G. J.,** GTP and cytosol stimulate phosphoinositide hydrolysis in isolated platelet membranes, *Biochem. Biophys. Res. Commun.,* 137, 801, 1986.

73. **Cockcroft, S. and Stutchfield, J.,** Effect of pertussis toxin and neomycin on G-protein-regulated polyphosphoinositide phosphodiesterase, *Biochem. J.,* 256, 343, 1988.

74. **Crouch, M. F. and Lapetina, E. G.,** A role for G$_i$ in control of thrombin receptor-phospholipase C coupling in human platelets, *J. Biol. Chem.,* 263, 3363, 1988.

75. **Grillone, L. R., Clark, M. A., Godfrey, R. W., Stassen, F., and Crooke, S. T.,** Vasopressin induces V$_1$ receptors to activate phosphatidylinositol- and phosphatidylcholine-specific phospholipase C and stimulates the release of arachidonic acid by at least two pathways in the smooth muscle cell line, A-10, *J. Biol. Chem.,* 263, 2658, 1988.

76. **Imboden, J. B., Shoback, D. M., Pattison, G., and Stobo, J. D.,** Cholera toxin inhibits the T-cell antigen receptor-mediated increase in inositol trisphosphate and cytoplasmic free calcium, *Proc. Natl. Acad. Sci. U.S.A.,* 83, 5673, 1986.

77. **Trimble, E. R., Bruzzone, R., Biden, T. J., Meehan, C. J., Andrea, D., and Merrifield, R. B.,** Secretin stimulates cyclic AMP and inositol trisphosphate production in rat pancreatic acinar tissue by two fully independent mechanisms, *Proc. Natl. Acad. Sci. U.S.A.,* 84, 3146, 1987.

78. **Low, W. W. Y. and Hughes, J.,** A novel cholera toxin-sensitive G-protein (G$_c$) regulating receptor-mediated phosphoinositide signalling in human pituitary clonal cells, *FEBS Lett.,* 220, 327, 1987.

79. **Masters, S. B., Martin, M. W., Harden, T. K., and Brown, J. H.,** Pertussis toxin does not inhibit muscarinic-receptor-mediated phosphoinositide hydrolysis of calcium mobilisation, *Biochem. J.,* 227, 933, 1985.

80. **Murayama, T. and Ui, M.,** Receptor-mediated inhibition of adenylate cyclase and stimulation of arachidonic acid release in 3T3 fibroblasts, *J. Biol. Chem.,* 260, 7226, 1985.

81. **Hepler, J. R. and Harden, T. K.,** Guanine nucleotide-dependent pertussis-toxin-insensitive stimulation of inositol phosphate formation by carbachol in a membrane preparation from human astrocytoma cells, *Biochem. J.,* 239, 141, 1986.

82. **Martin, T. F. J.**, Lipid hydrolysis by phosphoinositidase C: enzymology and regulation by receptors and guanine nucleotides, in *Inositol Lipids in Cell Signalling*, Michell, R. H., Drummond, A. H., and Downes, C. P., Eds., Academic Press, London, 1989.

83. **Scolnick, E. M., Papageorge, A. G., and Shih, T. Y.**, Guanine nucleotide binding activity as an assay for *src* protein of rat-derived murine sarcoma viruses, *Proc. Natl. Acad. Sci. U.S.A.*, 76, 5355, 1979.

84. **McGrath, J. P., Capon, D. J., Goeddel, D. V., and Levinson, A. D.**, Comparative biochemical properties of normal and activated human *ras* p21 proteins, *Nature*, 310, 644, 1984.

85. **Sweet, R. W., Yokoyama, S., Kamata, T., Feramisco, J. R., Rosenberg, M., and Gross, M.**, The product of *ras* is a GTPase and the T24 oncogenic mutant is deficient in this activity, *Nature*, 311, 273, 1984.

86. **Broek, D., Samiy, N., Fasano, O., Fujiyama, A., Tamanoi, F., Northup, J., and Wigler, M.**, Differential activation of yeast adenylate cyclase by wild-type and mutant RAS proteins, *Cell*, 41, 763, 1985.

87. **Beckner, S. K., Hattori, S., and Shih, T. Y.**, The *ras* oncogene product p21 is not a regulatory component of adenylate cyclase, *Nature*, 317, 71, 1985.

88. **Birchmeier, C., Broek, D., and Wigler, M.**, RAS proteins can induce meiosis in Xenopus oocytes, *Cell*, 43, 615, 1985.

89. **Fleischman, L. F., Chahwala, S. B., and Cantley, L.**, *ras*-transformed cells: altered levels of phosphatidylinositol-4,5-bisphosphate and catabolites, *Science*, 231, 407, 1986.

90. **Preiss, J., Loumis, C. R., Bishop, W. R., Stein, R., Niedel, J. E., and Bell, R. M.**, Quantitative measurement of sn-1,2-diacylglycerols present in platelets, hepatocytes and *ras*- and *cis*-transformed normal rat kidney cells, *J. Biol. Chem.*, 261, 8597, 1986.

91. **Marshall, C. J.**, Oncogenes and growth control, *Cell*, 49, 723, 1987.

92. **Wolfman, A. and Macara, I. G.**, Elevated levels of diacylglycerol and decreased phorbol ester sensitivity in *ras*-transformed fibroblasts, *Nature*, 325, 359, 1987.

93. **Bar-Sagi, D. and Feramisco, J. R.**, Induction of membrane ruffling and fluid-phase pinocytosis in quiescent fibroblasts by *ras* proteins, *Science*, 233, 1061, 1986.

94. **Alonso, T., Morgan, R. O., Marvizon, J. C., Zarbl, H., and Santos, E.**, Malignant transformation by *ras* and other oncogenes produces common alterations in inositol phospholipid signaling pathways, *Proc. Natl. Acad. Sci. U.S.A.*, 85, 4271, 1988.

95. **Kim, D., Lewis, D. L., Graziadel, L., Neer, E. J., Bar-Sagi, D., and Clapham, D. E.**, G-protein βγ-subunits activate the cardiac muscarinic K$^+$-channel via phospholipase A$_2$, *Nature*, 337, 557, 1989.

96. **Benjamin, C. W., Tarpley, W. G., and Gorman, R. T.**, Loss of platelet-derived growth factor-stimulated phospholipase activity in NIH-3T3 cells expressing the EJ-ras oncogene, *Proc. Natl. Acad. Sci. U.S.A.*, 84, 546, 1987.

97. **Parries, G., Hobel, R., and Racker, E.**, Opposing effects of *ras* oncogene on growth factor-stimulated phosphoinositide hydrolysis: desensitization to platelet-derived growth factor and enhanced sensitivity to bradykinin, *Proc. Natl. Acad. Sci. U.S.A.*, 84, 2648, 1987.

98. **Lacal, J. C., Moscat, J. and Aaronson, S. A.**, Novel source of 1,2-diacylglycerol elevated in cells transformed by Ha-*ras* oncogene, *Nature*, 330, 269, 1987.

99. **Kung, H. F., Smith, M. R., Bekesi, E., Manne, V., and Stacey, D. W.**, Reversal of transformed phenotype by monoclonal antibodies against Ha-*ras* p21 proteins, *Exp. Cell. Res.*, 162, 363, 1986.

100. **Yu, C.-L., Tsai, M. H., and Stacey, D.**, Cellular *ras* activity and phospholipid metabolism, *Cell*, 52, 63, 1988.

101. **Bennet, C. F. and Crooke, S. T.**, Purification and characterization of a phosphoinositide-specific phospholipase C from guinea pig uterus, *J. Biol. Chem.*, 262, 13789, 1987.

102. **Tsai, M.-H., Yu, C.-L., Wei, F.-S., and Stacey, D. W.**, The effect of GTPase activating protein upon *ras* is inhibited by mitogenically responsive lipids, *Science*, 243, 522, 1989.

103. **Coughlin, S. R., Escobedo, J. A., and Williams, L. T.**, Role of phosphatidylinositol kinase in PDGF receptor signal transduction, *Science*, 243, 1192, 1989.

104. **Berridge, M. B.**, Inositol trisphosphate and diacylglycerol: two interacting second messengers, *Annu. Rev. Biochem.*, 56, 159, 1987.

105. **Wollheim, C. B. and Biden, T. J.**, Second messenger function of inositol 1,4,5-trisphosphate, *J. Biol. Chem.*, 261, 8314, 1986.

106. **Thomas, A. P., Alexander, J., and Williamson, J. R.**, Relationship between inositol polyphosphoinositide production and the increase of cytosolic free Ca^{2+} induced by vasopressin in isolated hepatocytes, *J. Biol. Chem.*, 259, 5574, 1984.

107. **Reynolds, E. E. and Dubyak, G. R.**, Activation of calcium mobilization and calcium influx by alpha 1-adrenergic receptors in a smooth muscle cell line, *Biochem. Biophys. Res. Commun.*, 130, 627, 1985.

108. **Ramsdell, J. S. and Tashjian, A. H.**, Thryrotropin-releasing hormone elevation of inositol trisphosphate and cytosolic free calcium is independent of receptor number, *J. Biol. Chem.*, 261, 5301, 1986.

109. **Prescott, S. M. and Majerus, P. W.**, The fatty acid composition of phosphatidylinositol from thrombin-stimulated human platelets, *J. Biol. Chem.*, 256, 579, 1981.

110. **Downes, C. P., Mussat, M. C., and Michell, R. H.,** The inositol trisphosphate phosphomonoesterase of the human erythrocyte membrane, *Biochem. J.,* 203, 169, 1982.

111. **Connolly, T. M., Bross, T. E., and Majerus, P. W.,** Protein kinase C phosphorylates human platelet inositol trisphosphate 5-phosphomonoesterase, increasing the phosphatase activity, *J. Biol. Chem.,* 260, 7868, 1985.

112. **Connolly, T. M., Lawing, W. J., Jr., and Majerus, P. W.,** Isolation of a phosphomonoesterase from human platelets that specifically hydrolyzes the 5-phosphate of inositol 1,4,5-trisphosphate, *Cell,* 46, 951, 1986.

113. **Molina y Vedia, L. M. and Lapetina, E. G.,** Phorbol 12,13-dibutyrate and 1-oleyl-2-acetyldiacylglycerol stimulate inositol trisphosphate dephosphorylation in human platelets, *J. Biol. Chem.,* 261, 10493, 1986.

114. **Bloomquist, B. T., Shortridge, R. D., Schneuwly, S., Perdew, M., Montell, C., Steller, H., Rubin, G., and Pak, W. L.,** Isolation of a putative phospholipase C gene of Drosophila norpA, and its role in phototransduction, *Cell,* 54, 723, 1988.

115. **Baer, K. M. and Saibil, H. R.,** Light- and GTP-activated hydrolysis of phosphatidylinositol bisphosphate in squid photoreceptor membranes, *J. Biol. Chem.,* 263, 17, 1988.

116. **Devary, O., Heichal, O., Blumenfeld, A., Cassel, D., Suss, E., Barash, S., Rubenstein, C. T., Minke, B., and Selinger, Z.,** Coupling of photoexcited rhodopsin to inositol phospholipid hydrolysis in fly photoreceptors, *Proc. Natl. Acad. Sci. U.S.A.,* 84, 6939, 1987.

117. **Irvine, R. F., Letcher, A. J., Lander, D. J., and Downes, C. P.,** Inositol trisphosphates in carbachol-stimulated rat parotid glands, *Biochem. J.,* 223, 237, 1984.

118. **Irvine, R. F., Anggard, E. E., Letcher, A. J., and Downes, C. P.,** Metabolism of inositol 1,4,5-trisphosphate and inositol 1,3,4-trisphosphate in rat parotid glands, *Biochem. J.,* 229, 505, 1985.

119. **Heslop, J. P., Blakely, D. M., Brown, K. D., Irvine, R. F., and Berridge, M. J.,** Effects of bombesin and insulin on inositol (1,4,5)trisphosphate and inositol (1,3,4)trisphosphate formation in swiss 3T3 cells, *Cell,* 47, 703, 1986.

120. **Hawkins, P. T., Stephens, L., and Downes, C. P.,** Rapid formation of inositol(1,3,4,5)tetrakisphosphate and inositol (1,3,4)trisphosphate in rat parotid glands may both result indirectly from receptor-stimulated release of inositol(1,4,5) trisphosphate from phosphatidylinositol(4,5)bisphosphate, *Biochem. J.,* 238, 507, 1986.

121. **Vallejo, M., Jackson, T., Lightman, S., and Hanley, M. R.,** Occurrence and extracellular actions of inositol pentakis- and hexakisphosphate in mammalian brain, *Nature,* 330, 656, 1987.

122. **Morgan, R. O., Chang, J. P., and Catt, K. S.,** Novel aspects of gonadotropin-releasing hormone action on inositol polyphosphate metabolism in cultured pituitary gonadotrophs, *J. Biol. Chem.,* 262, 1166, 1987.

123. **Streb, H., Irvine, R. F., Berridge, M., and Schultz, I.,** Release of Ca^{2+} from a non-mitochondrial intracellular store of pancreatic acinar cells by inositol 1,4,5-trisphosphate, *Nature,* 306, 67, 1983.

124. **Ehrlich, B. E. and Watras, J.,** Inositol 1,4,5-trisphosphate activates a channel from smooth muscle sarcoplasmic reticulum, *Nature,* 336, 583, 1988.

125. **Vilven, J. and Coronado, R.,** Opening of dihydropyridine calcium channels in skeletal muscle membranes by inositol trisphosphate, *Nature,* 336, 587, 1988.

126. **Irvine, R. F. and Moor, R. M.,** Micro-injection of inositol 1,3,4,5-tetrakisphosphate activates sea urchin eggs, requires the presence of inositol trisphosphate, *Biochem. J.,* 240, 917, 1986.

127. **Suh, P.-G., Ryu, S. H., Moon, K. H., Suh, H. W., and Rhee, S. G.,** Cloning and sequence of multiple forms of phospholipase C, *Cell,* 54, 161, 1988.

128. **Katan, M., Kriz, R. W., Totty, N., Philp, R., Meldrum, E., Aldape, R. A., Knopf, J. L., and Parker, P. J.,** Determination of the primary structure of PLC-154 demonstrates diversity of phosphoinositide-specific phospholipase C activities, *Cell,* 54, 171, 1988.

129. **Teisseire, B., Ropars, C., Villereal, M. C., and Nicolau, C.,** Long-term physiological effects of enhanced O_2 release by inositol hexaphosphate-loaded erythrocytes, *Proc. Natl. Acad. Sci. U.S.A.,* 84, 6894, 1987.

130. **Kikkawa, U. and Nishizuka, Y.,** The role of protein kinase C in transmembrane signalling. *Annu. Rev. Cell Biol.,* 2, 149, 1986.

131. **Rando, R. R.,** Regulation of protein kinase C activity by lipids, *FASEB J.,* 2, 2348, 1988.

132. **Boni, L. T. and Rando, R. R.,** The nature of protein kinase C activation by physically defined phospholipid vesicles and diacylglycerol, *J. Biol. Chem.,* 260, 10819, 1985.

133. **Mori, T., Takai, Y., Yu, B., Takahashi, J., Nishizuka, Y., and Fujikura, T.,** Specificity of the fatty acyl-moieties of diacylglycerol for the activation of calcium-activated, phospholipid-dependent protein kinase, *J. Biochem.,* 91, 427, 1982.

134. **Ganong, B. R., Loomis, C. R., Hannun, Y. A., and Bell, R. M.,** Specificity and mechanism of protein kinase C activation by ss-1,2-diacylglycerols, *Proc. Natl. Acad. Sci. U.S.A.,* 83, 1184, 1986.

135. **Heymans, F., DaSilva, C., Marrec, N., Godfroid, J. J., and Castagna, M.,** Alkyl analogs of diacylglycerol as activators of protein kinase C, *FEBS Lett.,* 218, 35, 1987.

136. **Coussens, L., Parker, P. J., Rhee, L., Yang-Feng, T. L., Chen, E., Waterfield, M. D., Francke, U., and Ullrich, A.,** Multiple, distinct forms of bovine and human protein kinase C suggest diversity in cellular signalling pathways, *Science,* 233, 859, 1986.

137. Knopf, J. L., Lee, M.-H., Sultzman, L. A., Kriz, R. W., Loomis, C. R., Hewick, R. M., and Bell, R. M., Cloning and expression of multiple protein kinase C cDNAs, *Cell*, 46, 491, 1986.

138. Makowske, M., Birnbaum, M. J., Ballester, B., and Rosen, O. M., A cDNA encoding PKC identifies two species of mRNA in brain and GH3 cells, *J. Biol. Chem.*, 261, 13389, 1986.

139. Parker, P. J., Coussens, L., Totty, N., Rhee, L., Young, S., Chen, E., Stabel, S., Waterfield, M. D., and Ulrich, A., The complete primary structure of protein kinase C—the major phorbol ester receptor, *Science*, 233, 853, 1987.

140. Ohno, S., Kawasaki, H., Imajoh, S., Suzuki, K., Inagaki, M., Yokohura, H., Sakoh, T., and Hidaka, H., Tissue-specific expression of three distinct types of rabbit protein kinase C, *Nature*, 325, 161, 1987.

141. Ono, Y., Kikkawa, U., Ogita, K., Tomoko, F., Kurokawa, T., Asaoka, Y., Sekiguchi, K., Ase, K., Igarashi, K., and Nishizuka, Y., Expression and properties of two types of protein kinase C: alternative splicing from a single gene, *Science*, 236, 1116, 1987.

142. Brandt, S. J., Niedel, J. E., Bell, R. M., and Young, W. S., III, Distinct patterns of expression of different protein kinase C mRNAs in rat tissues, *Cell*, 49, 57, 1987.

143. Housey, G. M., Johnson, M. D., Hsiao, W. L. W., O'Brian, C. A., Murphy, J. P., and Weinstein, I. B., Isolation of cDNA clones encoding protein kinase C: evidence for a protein kinase C-related gene family, *Proc. Natl. Acad. Sci. U.S.A.*, 84, 1065, 1987.

144. Ono, Y., Fujii, T., Ogita, K., Kikkawa, U., Igarashi, K., and Nishizuka, Y., The structure, expression, and properties of additional members of the protein kinase C family, *J. Biol. Chem.*, 263, 6927, 1988.

145. Ohno, S., Akita, Y., Konno, Y., Imajoh, S., and Suzuki, K., A novel phorbol ester receptor/protein kinase, nPKC, distinctly related to the protein kinase C family, *Cell*, 53, 731, 1988.

146. Castagna, M., Takai, Y., Kaibuchi, K., Sano, K., Kikkawa, U., and Nishizuka, Y., Direct activation of calcium-activated, phospholipid-dependent protein kinase by tumor-promoting phorbol esters, *J. Biol. Chem.*, 257, 7847, 1982.

147. Housey, G. M., Johnson, M. D., Hsiao, W. L. W., O'Brian, C. A., Murphy, J. P., Kirschmeier, P., and Weinstein, I. B., Overproduction of protein kinase C causes disordered growth control in fibroblasts, *Cell*, 52, 343, 1988.

148. Pasti, G., Lacal, J. C., Warren, B. S., Aaronson, S. A., and Blumberg, P. M., Loss of mouse fibroblast cell response to phorbol esters restored by microinjected protein kinase C, *Nature*, 325, 161, 1987.

149. Kamata, T., Sullivan, N. F., and Wooten, M. W., Reduced protein kinase C activity in a ras-resistant cell line derived from Ki-MSV transformed cells, *Oncogene*, 1, 19, 1987.

150. Farooqui, A. A., Farooqui, T., Yates, A. J., and Horrocks, L. A., Regulation of protein kinase C activity by various lipids, *Neurochem. Res.*, 13, 499, 1988.

151. Gomez-Cambronero, J., Molski, T. F. P., Becker, E. L., and Sha'afi, R. I., The diacylglycerol kinase inhibitor R59022 potentiates superoxide production but not the secretion induced by fMet-Leu-Phe: effects of leupeptin and the protein kinase C inhibitor H-7, *Biochem. Biophys. Res. Commun.*, 148, 38, 1987.

152. Parker, J., Daniel, L. W., and Waite, M., Evidence of protein kinase C involvement in phorbol-diester-stimulated arachidonic acid release and prostaglandin synthesis, *J. Biol. Chem.*, 262, 5385, 1987.

153. Miller, R. J., Protein kinase C: a key regulator of neuronal excitability, *Trends Neurosci.*, 9, 538, 1986.

154. Wright, T. M., Hoffman, R. D., Nishijima, J., Jakoi, L., Snyderman, R., and Shin, H. S., Leukocyte chemoattraction by 1,2-diacylglycerol, *Proc. Natl. Acad. Sci. U.S.A.*, 85, 1869, 1988.

155. Strälfors, P., Insulin stimulation of glucose uptake can be mediated by diacylglycerol in adipocytes, *Nature*, 335, 554, 1988.

156. Yamakawa, A. and Takenawa, T., Purification of membrane-bound phosphatidylinositol kinase from rat brain, *J. Biol. Chem.*, 263, 17555, 1988.

157. Fukui, T., Lutz, R. J., and Lowenstein, J. M., Purification of a phospholipase C from rat liver cytosol that acts on phosphatidylinositol 4,5-bisphosphate and phosphatidylinositol 4-phosphate, *J. Biol. Chem.*, 263, 17730, 1988.

158. Fukami, K., Matsuoka, K., and Nakanishi, O., Antibody to phosphatidylinositol 4,5-bisphosphate inhibits oncogene-induced mitogenesis, *Proc. Natl. Acad. Sci. U.S.A.*, 85, 9057, 1988.

159. Baukal, A. J., Guillemette, G., Rubin, R., Spat, A., and Catt, K. J., Binding sites for inositol trisphosphate in the bovine adrenal cortex, *Biochem. Biophys. Res. Commun.*, 133, 532, 1985.

160. Spat, A., Fabiato, A., and Rubin, R. P., Binding of inositol trisphosphate by a liver microsomal fraction, *Biochem. J.*, 233, 929, 1986.

161. Worley, P. F., Baraban, J. M., Colvin, J. S., and Snyder, S. H., Inositol trisphosphate receptor localization in brain: variable stoichiometry with protein kinase C, *Nature*, 325, 159, 1987.

162. Supattapone, S., Worley, P. F., Baraban, J. M., and Snyder, S. H., Solubilization, purification, and characterization of an inositol trisphosphate receptor, *J. Biol. Chem.*, 263, 1530, 1988.

163. Supattapone, S., Danoff, S. K., Theibert, A., Joseph, S. K., Steiner, J., and Snyder, S. H., Cyclic AMP-dependent phosphorylation of a brain inositol trisphosphate receptor decreases its release of calcium, *J. Biol. Chem.*, 85, 8747, 1988.

164. Tabuse, Y., Nishiwaki, K., and Miwa, J., Mutations in a protein kinase C homolog confer phorbol ester resistance on *Caenorhabditis elegans*, *Science*, 243, 1713, 1989.

165. **Hockberger, P., Toselli, M., Swandulla, D., and Lux, H. D.,** A diacylglycerol analogue reduces neuronal calcium currents independently of protein kinase C activation, *Nature,* 338, 340, 1989.
166. **Joseph, S. K., Hansen, C. A., and Williamson, J. R.,** Inositol 1,3,4,5-tetrakisphosphate increases the duration of the inositol, 1,4,5-trisphosphate-mediated Ca^{2+} transient, *FEBS Lett.,* 219, 125, 1987.
167. **Berg, I., Guse, A. H., and Gercken, G.,** Carbamoylcholine-induced accumulation of inositol mono-, bis-, tris- and tetrakisphosphates in isolated myocytes from adult rats, *Biochim. Biophys. Acta,* 1010, 100, 1989.
168. **Baker, M. E.,** Similarity between phospholipase C and the regulatory domain of protein kinase C, *Mol. Cell. Endocrinol.,* 61, 129, 1989.
169. **Wahl, M. I., Nishibe, S., Suh, P. G., Rhee, S. G., and Carpenter, G.,** Epidermal growth factor stimulates tyrosine phosphorylation of phospholipase C-II independently of internalization and extracellular Ca^{2+}, *Proc. Natl. Acad. Sci. U.S.A.,* 86, 1568, 1989.
170. **Rhee, S. G., Suh, P. G., Ryu, S. H., and Lee, S. Y.,** Studies of inositol phospholipid-specific phospholipase C, *Science,* 244, 546, 1989.
171. **Berridge, M. J. and Galione, A.,** Cytosolic calcium oscillations, *FASEB J.,* 2, 3074, 1988.
172. **Wakui, M., Potter, B. V. L., and Petersen, O. H.,** Pulsatile intracellular calcium release does not depend on fluctuations in inositol trisphosphate concentration, *Nature,* 339, 317, 1989.
173. **Schaeffer, E., Smith, D., Mardon, G., Quinn, W., and Zuker, C.,** Isolation and characterization of 2 new Drosophila protein kinase-C genes, including one specifically expressed in photoreceptor cells, *Cell,* 57, 403, 1989.
174. **Ling, L. E., Schulz, J. T., and Cantley, L. C.,** Characterization and purification of membrane-associated phosphatidylinositol-4-phosphate kinase from human red blood cells, *J. Biol. Chem.,* 264, 5080, 1989.
175. **Smith, C. D. and Chang, K. Y.,** Regulation of brain phosphatidylinositol-4-phosphate kinase by GTP analogues, *J. Biol. Chem.,* 264, 3206, 1989.
176. **Kato, H., Uno, I., Ishikawa, T., and Takenawa, T.,** Activation of phosphatidylinositol kinase and phosphatidylinositol-4-phosphate kinase by cAMP in Saccharomyces cerevisiae, *J. Biol. Chem.,* 264, 3116, 1989.
177. **Boyer, J. L., Downes, C. P., and Harden, T. K.,** Kinetics of activation of phospholipase C by P_{2y} purinergic receptor agonists and guanine nucleotides, *J. Biol. Chem.,* 264, 884, 1989.
178. **Smith, M. R., Ryu, S. H., Suhm, P. G., and Kung, H. F.,** S-phase induction and transformation of quiescent NIH 3T3 cells by microinjection of phospholipase C, *Proc. Natl. Acad. Sci. U.S.A.,* 86, 3659, 1989.
179. **Balla, T., Hunyady, L., Bankal, A. J., and Catt, K. J.,** Structures and metabolism of inositol tetrakisphosphates and inositol pentakisphosphate in bovine adrenal glomerulosa cells, *J. Biol. Chem.,* 264, 9386, 1989.
180. **Mauger, J. P., Claret, M., Pietri, F., and Hilly, M.,** Hormonal regulation of inositol 1,4,5-trisphosphate receptor in rat liver, *J. Biol. Chem.,* 264, 8821, 1989.
181. **Whitman, M., Downes, C. P., Keeler, M., Keeler, T., and Cantley, L.,** Type I phosphatidylinositol kinase makes a novel inositol phospholipid, phosphatidylinositol-3-phosphate, *Nature,* 332, 644, 1988.
182. **Lips, D. L., Majerus, P. W., Gorga, F. R., Young, A. T., and Benjamin, T. L.,** Phosphatidylinositol 3-phosphate is present in normal and transformed fibroblasts and is resistant to hydrolysis by bovine brain phospholipase C II, *J. Biol. Chem.,* 264, 8759, 1989.
183. **Ross, C. A., MacCumber, M. W., Glatt, C. E., and Snyder, S. H.,** Brain phospholipase C isozymes: differential mRNA localizations by *in situ* hybridization, *Proc. Natl. Acad. Sci. U.S.A.,* 86, 2923, 1989.
184. **Nishibe, S., Wahl, M. I., Rhee, S. G., and Carpenter, G.,** Tyrosine phosphorylation of phospholipase C-II *in vitro* by the epidermal growth factor, *J. Biol. Chem.,* 264, 10335, 1989.
185. **Ashkenazi, A., Peralta, E. G., Winslow, J. W., Ramachandran, J., Capon, D. J.,** Functionally distinct G proteins selectively couple different receptors to PI hydrolysis in the same cell, *Cell,* 56, 487, 1989.
186. **Auger, K. R., Serunian, L. A., Soltoff, S. P., Libby, P., and Cantley, L. C.,** PCGF-dependent tyrosine phosphorylation stimulates production of novel polyphosphoinositides in intact cells, *Cell,* 57, 167, 1989.
187. **Siegel, D. P., Banschbach, J., Alford, D., Ellens, H., Lis, L. J., Quinn, P. J., Yeagle, P. L., and Bentz, J.,** Physiological levels of diacylglycerol in phospholipid membranes induce membrane fusion and stabilize inverted phases, *Biochemistry,* 28, 3703, 1989.
188. **Margolis, B., Rhee, S. G., Felder, S., Mervic, M., Lyall, R., Levitzki, A., Ullrich, A., Zilberstein, A., and Schlessinger, J.,** EGF induces tyrosine phosphorylation of phospholipase C-II: a potential mechanism for EGF receptor signaling, *Cell,* 57, 1101, 1989.
189. **Meisenhelder, J., Suh, P. G., Rhee, S. G., and Hunter, T.,** Phospholipase C-τ is a substrate for the PDGF and EGF receptor protein-tyrosine kinases in vivo and in vitro, *Cell,* 57, 1109, 1989.
190. **Macara, I. G.,** Elevated phosphocholine concentration in *ras*-transformed NIH 3T3 cells arises from increased choline kinase activity, not from phosphatidylcholine breakdown, *Mol. Cell. Biol.,* 9, 325, 1989.
191. **Johnson, R. M., Wasilenko, W. J., Mattingly, R. R., Weber, M. J., and Garrison, J. C.,** Fibroblasts transformed with v-*src* show enhanced formation of an inositol tetrakisphosphate, *Science,* 246, 121, 1989.

192. **Somlyo, A. P., Walker, J. W., Goldman, Y. E., Trentham, D. R., Kobayashi, F. R. S. S., Kitazawa, T., and Somlyo, A. V.**, Inositol trisphosphate and muscle contraction, *Philos. Trans. R. Soc. London B*, 320, 399, 1988.

193. **Stephens, L. R., Hawkins, P. T., Morris, A. J., and Downes, C. P.**, L-*myo*inositol 1,4,5,6-tetrakisphosphate 3-hydroxyl kinase, *Biochem. J.*, 249, 271, 1988.

194. **Stephens, L. R., Hawkins, P. T., Carter, N., Chahwala, S. B., Morris, A. J., Whetton, A. D., and Downes, C. P.**, L-*myo*inositol 1,4,5,6-tetrakisphosphate is present in both mammalian and avian cells, *Biochem. J.*, 249, 271, 1988.

195. **Irvine, R. F. and Moor, R. M.**, Inositol(1,3,4,5)tetrakisphosphate-induced activation of sea urchin eggs requires the presence of inositol trisphosphate, *Biochem. Biophys. Res. Commun.*, 146, 284, 1987.

196. **Crossley, I., Swann, K., Chambers, E., and Whitaker, M.**, Activation of sea urchin eggs is independent of external calcium ions, *Biochem. J.*, 252, 257, 1988.

197. **Morris, A. P., Gallacher, D. V., Irvine, R. F., and Petersen, O. H.**, Synergism of inositol 1,3,4,5-tetrakisphosphate with inositol 1,4,5-trisphosphate in mimicking muscarinic receptor activation of Ca^{2+}-dependent K^+ channels, *Nature*, 330, 653, 1987.

198. **Spät, A., Lukacs, G. L., Eberhardt, I., Kiesel, L., and Runnebaum, B.**, Binding of inositol phosphates and induction of Ca^{2+} release from pituitary microsomal fractions, *Biochem. J.*, 244, 493, 1987.

199. **Irvine, R. F., Moor, R. M., Pollock, W. K., Smith, P. M., and Wreggett, K. A.**, Inositol phosphates: proliferation, metabolism and function, *Philos. Trans. R. Soc. London B*, 320, 281, 1988.

200. **Sagi-Eisenberg, R.**, GTP-binding proteins as possible targets for protein kinase C action, *Trends Biochem. Sci.*, 14, 355, 1989.

201. **Ferris, C. D., Huganir, R. L., Supattapone, S., and Snyder, S. H.**, Purified inositol 1,4,5-trisphosphate receptor mediates calcium flux in reconstituted lipid vesicles, *Nature*, 342, 87, 1989.

202. **Furuichi, T., Yoshikawa, S., Miyawaki, A., Wada, K., Maeda, N., and Mikoshiba, K.**, Primary structure and functional expression of the inositol 1,4,5-trisphosphate-binding protein P_{400}, *Nature*, 342, 32, 1989.

203. **Mignery, G. A., Südhof, T. C., Takei, K., and De Camilli, P.**, Putative receptor for inositol 1,4,5-trisphosphate similar to ryanodine receptor, *Nature*, 342, 192, 1989.

204. **Balla, T., Baukal, A. J., Hunyady, L., and Catt, K. J.**, Agonist-induced regulation of inositol tetrakisphosphate isomers and inositol pentakisphosphate in adrenal glomerulosa cells, *J. Biol. Chem.*, 264, 13605, 1989.

205. **Price, B. D., Morris, J. D. H., Marshall, C. J., and Hall, A.**, Stimulation of phosphatidylcholine hydrolysis, diacylglycerol release, and arachidonic acid production by oncogenic RAS is a consequence of protein kinase-C activation, *J. Biol. Chem.*, 264, 16638, 1989.

206. **Varticovski, L., Druker, B., Morrison, D., Cantley, L., and Roberts, T.**, The colony stimulating factor-1 receptor associates with and activates phosphatidylinositol-3-kinase, *Nature*, 342, 699, 1989.

207. **Megidish, T. and Mazurek, N.**, A mutant protein kinase C that can transform fibroblasts, *Nature*, 342, 807, 1989.

Chapter 8

ROLE OF GLYCOSYL PHOSPHATIDYLINOSITOLS IN INSULIN SIGNALING

José M. Mato and Isabel Varela

TABLE OF CONTENTS

I. INTRODUCTION

Insulin is one of the best-studied hormones. The main actions of insulin are to stimulate the synthesis of glycogen, lipids, and proteins, through modulation of the metabolic pathways implicated in these processes, and the stimulation of cell growth. The physiological effects of insulin include stimulation of the uptake of glucose, amino acids, and ions; regulation of the state of serine and threonine phosphorylation of a variety of proteins and rate-limiting enzymes like glycogen phosphorylase, hormone-sensitive lipase, and ATP citrate lyase; and regulation of the expression of the genes for several regulatory enzymes.[1] All these effects, whose chronology varies from seconds to hours, are initiated by the interaction of the hormone with the insulin receptor, an integral membrane glycoprotein composed of two α (Mr, about 130 kDa) and two β (95 kDa) subunits. The α subunits bind insulin and are linked by disulphide bonds to each other and to the β subunits. Following insulin binding, the β subunits are rapidly autophosphorylated, predominantly at tyrosine residues.[2] The insulin receptor is an insulin-dependent protein tyrosine kinase.[3] Direct evidence indicating that the insulin receptor protein tyrosine kinase activity is necessary for many of the actions of insulin, including the stimulations of S6 kinase, deoxyglucose uptake, glycogen synthesis, thymidine incorporation into DNA, and receptor down-regulation, has been obtained by site-directed mutagenesis of the insulin receptor cDNA as well as with monoclonal antibodies to the insulin receptor kinase domain.[2,4,5] How receptor autophosphorylation is coupled to the modulation of processes like the synthesis of glycogen, lipids, and proteins is not well understood. One hypothesis suggests that the receptor kinase activity catalyzes the phosphorylation of cellular protein substrates. A second hypothesis suggests that autophosphorylation of the receptor would lead to variations in its interactions with other membrane components. In both cases, the interaction of insulin with its receptor would lead to the generation of biochemical signals involved in insulin action. Recently, evidence has been provided which suggests that inositol phospho-oligosaccharides (or inositol phospho-glycans) might be one of the biochemical signals involved in insulin signaling.[6]

II. STRUCTURE OF THE INSULIN-SENSITIVE GLYCOSYL PHOSPHATIDYLINOSITOL

Insulin has been reported to promote the phosphodiesteratic hydrolysis of a novel glycosyl phosphatidylinositol (glycosyl-PI) in a number of cells, including intact murine BC3H1 myocytes,[7] H35 hepatoma cells,[8] T lymphocytes,[9] and CHO cells.[10] These insulin-sensitive glycosyl-PIs have features in common with the glycosyl-PI anchor of a number of membrane proteins (also see Chapter 5).[11,12] The similarities include: (1) the finding that the insulin-sensitive glycosyl-PI is hydrolyzed by the phosphatidylinositol-specific phospholipase C (PI-PlC) from *Staphylococcus aureus, Bacillus cereus,* or *B. thuringiensis,*[7-10] (2) the observation that purified glycosyl-PI can be cleaved by nitrous acid deamination with generation of phosphatidylinositol, indicating the presence of inositol monophosphate linked to non-N-acetylated glucosamine,[8,13,14] and (3) the finding that the diacylglycerol moiety of the phospholipid contains saturated fatty acids (mainly palmitate and myristate).[7,8] In H35 cells, the isolated glycosyl-PI has a 1-alkyl-2-acylglycerol structure.[8] In addition, the polar headgroup of the insulin-sensitive glycosyl-PI isolated from rat liver membranes or H35 hepatoma cells has been shown to contain galactose (about four residues per residue of inositol) and up to three additional phosphates,[13,15] probably linked to a galactose residue through a monoester bond.[15] Both *myo*-inositol and *chiro*-inositol have been detected in the purified glycolipid from H35 hepatoma cells and rat liver membranes.[13,14] *Chiro*-inositol has been detected in the hydrophobic membrane-anchoring domain of *Torpedo* acetylcholinesterase[16,17] and of human placental alkaline phosphatase.[17] The level of *chiro*-inositol in the above examples

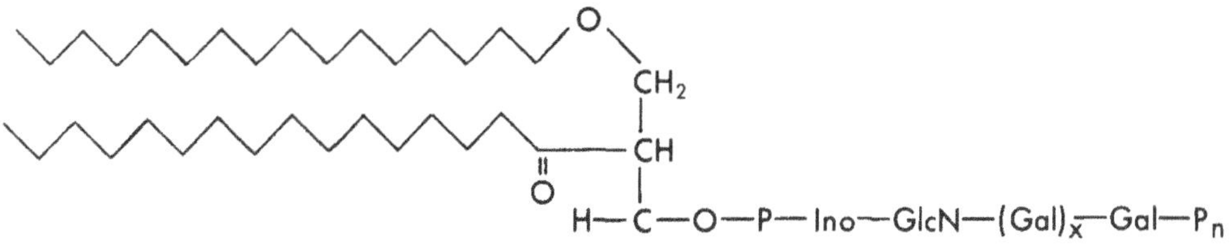

FIGURE 1.   Model structure of the insulin-sensitive glycosyl-PI. Ino, *chiro/myo*-inositol; GlcN, non-N-acetylated glucosamine; Gal, galactose. The number of galactose residues has been tentatively determined as three or four. The glycophospholipid exists in several forms, which differ in the number of phosphate residues (tentatively, from one to three) associated with one residue of galactose. These phosphates form monoester bonds.

varied from 50 to 3% of the amount of *myo*-inositol.[16,17] The finding of both *myo*- and *chiro*-inositol suggests the presence of two forms of the same insulin-sensitive glycosyl-PI with different inositol isomers. The purified insulin-sensitive glycosyl-PI is also sensitive to treatment with β-galactosidase,[18] indicating the presence of galactose residues with β-glycosidic bonds. Another characteristic of this glycosyl-PI is the absence of amino acids and ethanolamine.[8,13] The sensitivity to bacterial PI-PlC of insulin-sensitive glycosyl-PIs isolated from H35 hepatoma cells, mice T lymphocytes, and CHO cells is different,[7,8] and incomplete hydrolysis has been observed with the glycosyl-PI isolated from mice T lymphocytes,[8] CHO cells,[10] and in Rat-1 cells.[82] The incomplete hydrolysis by PI-PlC may suggest the presence of a set of closely related glycosyl-PI molecules with different degrees of sensitivity to enzymatic cleavage. Figure 1 shows a model structure of the insulin-sensitive glycosyl-PI.

Larner's laboratory has reported the carbohydrate composition of two "insulin mediators" purified from rat liver which presumably are the polar headgroups of two glycosyl-PIs. One of the purified substances contained *chiro*-inositol, galactosamine, and mannose, and the second substance contained *myo*-inositol, glucosamine, and galactose.[14] These results suggest the existence of heterogeneity between insulin-sensitive glycosyl-PIs from different cell types and even within one cell type. In fact, up to three different polar headgroups have been separated from rat liver, using anion chromatography, which presumably differ in the number of phosphate groups.[7,15] A family of glycosyl-PIs has been isolated from *Leishmania major*.[19] The compositional analyses indicate that the glycan chain of the glycosyl-PIs have between four and ten saccharide residues, and are consistent with each of the glycosyl-PIs containing one residue of *myo*-inositol and non-N-acetylated glucosamine and variable amounts of galactose (one to four residues), mannose (two to four residues), glucose (0 to one residues), and phosphate (one to three residues), and an absence of ethanolamine.[19] These results indicate that glycosyl-PIs are found in many different cell types and that they probably perform a variety of biological functions.

The majority of the insulin-sensitive glycosyl-PI is present in the plasma membrane (where it represents about 0.5% of the total rat liver phospholipids) and, as with most glycolipids, the bulk of it seems to be present at the outer surface of rat hepatocytes[20] and adipocytes.[18] These results have been obtained by reacting intact and broken cells with [¹⁴C] isethyonyl acetimidate (see Chapter 4) and comparing the amount of labeled lipid under both conditions, or by incubating [³H] glucosamine-labeled cells (both intact and broken) with β-galactosidase or PI-PlC and comparing the amount of glycosyl-PI that remains after both treatments. In *L. major*, two glycosyl-PIs have been specifically labeled on intact viable promastigotes with NaB[³H]₄ and galactose oxidase, indicating that these lipids are also expressed on the outside of the plasma membrane.[19] The observation that the majority of the glycosyl-PI is at the outer cell surface indicates that in response to the addition of insulin, the inositol phospho-oligosaccharide (POS) that forms the polar headgroup of the glycolipid will be generated extracellularly. In connection with this, it has been reported that the extracellular concentration of POS increased rapidly after the addition of insulin to BC3H1 myocytes.[14] Moreover, a POS-like factor which mimics some of the actions of insulin has been reported to be constitutively released to the medium by Reuber hepatoma cells.[22] Furthermore, as a result of the phosphodiesteratic hydrolysis of glycosyl-PI, diacylglycerol is also formed at the outer leaflet of the membrane bilayer. Since the rate of transbilayer movement of diacylglycerol is slow (see Chapter 3), at least in comparison to the rate of the initial actions of insulin, the physiological role, if any, of the diacylglycerol generated upon glycosyl-PI hydrolysis might be confined to the outer surface of the cell, and not necessarily be coupled to protein kinase C translocation and activation (see Chapter 7).

The absence of ethanolamine in the molecule of glycosyl-PI suggests that this glycolipid is not the immediate precursor for protein anchoring, since this residue is necessary to form the amide bond to the carboxyl-terminal amino acid of the protein (see Chapter 5). Moreover, the putative precursor for glycosyl-PI anchoring is most likely to be present in the endoplasmic reticulum, where protein synthesis takes place,[23,24] whereas the insulin-sensitive glycosyl-PI is present at the plasma membrane.[20] However, these glycosyl-PIs might be the products of the cleavage of proteins that are PI anchored. Other possible functions for these glycosyl-PIs, in addition to a role in cellular signaling (see below), might be the regulation of the activity of certain enzymes, as has been suggested for the lipophosphoglycan (LPG) and glycosyl-PI antigens in parasites.[25]

III. INSULIN-DEPENDENT GLYCOSYL-PI HYDROLYSIS

As mentioned above, the addition of insulin to a number of target cells stimulates the hydrolysis of glycosyl-PI. This process is dose-dependent and can be observed at physiological concentrations of the hormone (0.1 to 1.0 nM).[8,10] The time course of glycosyl-PI hydrolysis is fast, and can be detected within 1 min of the addition of insulin.[7-10] As previously mentioned, the protein tyrosine kinase activity of the insulin receptor is necessary for many of the biological actions of insulin.[2] The requirement of this insulin-receptor tyrosine kinase activity in insulin-dependent glycosyl-PI hydrolysis has been evaluated by analyzing glycosyl-PI hydrolysis in stably transfected CHO cell lines expressing either the wild-type human insulin receptor (CHO wt) or a receptor mutated at the ATP binding site (Lys 1018 → Ala) that lacks tyrosine kinase activity (CHO mut).[4,5] The parental CHO cells expressed about 1000 receptors per cell, whereas the two transfected cell lines expressed approximately 40,000 high-affinity human insulin receptors per cell.[4,5,10] Cells bearing normal human receptors (CHO wt) hydrolyzed up to 70% of the glycosyl-PI within 2 min of the addition of 0.1 nM insulin.[10] Thus, as in other cells, normal human insulin receptors are coupled to the glycosyl-PI hydrolyzing system of the host CHO cell. The responses of both CHO and CHO mut cells to insulin were similar to each other and markedly different from CHO wt. Thus, the parental CHO cells hydrolyzed about 30% of the glycosyl-PI upon insulin addition, and the CHO mut hydrolyzed about 20% of the glycolipid.[7] These results strongly suggest that the mutated receptor is unable to transduce the effect of insulin on glycosyl-PI hydrolysis. The 30% hydrolysis in CHO mut cells may be due to the contribution of the functional endogenous insulin receptor in these CHO cells or to incomplete hydrolysis produced by the defective insulin receptors. The effect of insulin on glycosyl-PI hydrolysis was dose dependent. In CHO wt cells, half-maximal hydrolysis was obtained at about 0.01 nM insulin.[10] These cells hydrolyzed about 70% of their glycosyl-PI content in response to saturating doses of insulin. In parental CHO cells and CHO mut cells, half-maximal glycosyl-PI hydrolysis occurred at 0.01 nM and 0.1 nM, respectively.[10] The maximal effect of insulin on glycosyl-PI hydrolysis was markedly reduced, and only about 30 and 20% of the glycolipid was hydrolyzed in response to saturating doses of insulin in CHO cells and CHO mut cells, respectively.[10] Thus, unlike some of the other effects of insulin (e.g., activation of glycogen synthase and S6 kinase), cells bearing few receptors or kinase-deficient receptors do not achieve the same level of glycosyl-PI hydrolysis as CHO wt cells. This may reflect the requirement of a quantum of insulin receptor kinase activity that can only be generated in cells with a relatively high number of normal insulin receptors.

The mechanism by which the insulin-receptor tyrosine kinase activity stimulates the phospholipase C responsible for glycosyl-PI hydrolysis is not known. The addition of insulin to myocytes and adipocytes has been shown to induce the release of alkaline phosphatase[21] and lipoprotein lipase,[26] respectively, two enzymes which are attached to the membrane through a glycosyl-PI anchor (see Chapter 5), indicating that these cells contain an insulin-sensitive phospholipase C whose active site is at the outer cell surface. In the homogenates of fat cells, it has been shown that insulin induced the stimulation of a phospholipase C acting on phosphatidylinositol.[35] Moreover, a phospholipase C acting on glycosyl-PI anchors has been partially purified from rat liver.[27] However, whether this enzyme is responsible for the hydrolysis of glycosyl-PI after the addition of insulin is unknown. Recently, evidence has been given supporting the view that EGF and PDGF stimulation of the receptor tyrosine kinase activity might activate PIP_2 hydrolysis through tyrosine phosphorylation of PI-PlC II (see Chapter 7).[28,29,64,65] Whether stimulation of the insulin-receptor protein tyrosine kinase activity triggers glycosyl-PI hydrolysis through tyrosine phosphorylation of a specific PI-PlC remains to be determined. The glycosyl-PI-specific phospholipase C of *Trypanosome brucei* has been localized in intracellular organelles, but was absent from plasma membranes.[62] If the localization of the insulin-stimulated glycosyl-PI phospholipase C is also in intracellular organelles, this raises the possibility that the enzyme is translocated to the plasma membrane in response to insulin when the hydrolysis of glycosyl-PI is needed.

A number of experiments indicate that there is a correlation between the number of insulin receptors and the levels of glycosyl-PI. Thus, in the parental CHO cells and in the CHO mut cells bearing the mutated human insulin receptor that lacks protein tyrosine kinase activity, the amount of glycosyl-PI is about threefold lower than that found in cells bearing the normal insulin receptor.[10] A correlation of glycosyl-PI levels with the induction of insulin receptors has also been found in T lymphocytes,[9] and there is a correlation of glycosyl-PI levels with the insulin-induced reduction of their own receptors (down-regulation) in rat hepatocytes.[30] Moreover, it has been shown that insulin resistance caused by dexamethasone administration is accompanied by a marked reduction in the hepatocyte levels of glycosyl-PI as well as by a blockade of its hydrolysis in response to insulin.[31] In contrast, bilateral adrenalectomy raised the cellular content of glycosyl-PI and enhanced its hydrolysis in response to insulin. Simultaneous with these changes, dexamethasone treatment reduced, and adrenalectomy enhanced, the total number of insulin receptors in liver cells.[31] It is important to note that the administration of dexamethasone almost completely abolished the stimulatory effect of insulin on hepatocyte glycogen synthesis.[31] While the causes of these changes in glycosyl-PI levels in diabetes are not understood, the correlation of the changes in the number of biologically active insulin receptors present in the cell and the levels of glycosyl-PI seems to be an important new direction in this area and might be related to the mechanism of insulin resistance.

The possibility that other receptors with protein tyrosine kinase activity might be coupled to glycosyl-PI hydrolysis has been investigated by various laboratories. EGF (100 ng/ml) and IGF-I (50 ng/ml) have been reported to stimulate the hydrolysis of the insulin-sensitive glycosyl-PI in a time-dependent manner in BC3H1 myocytes.[32] In CHO cells, IGF-I (15 nM) has a small effect on glycosyl-PI hydrolysis (10%) which increases to 20% in cells transfected with the human IGF-I receptor and that exhibit about four times more receptors than the parental CHO cells.[10] Although suggestive, the small effect of IGF-I on glycosyl-PI hydrolysis in CHO raises the possibility that it may be mediated by the insulin receptor. In PC-12 cells, NGF (50 ng/ml) stimulates the hydrolysis of glycosyl-PI.[33] Although tyrosine phosphorylation has been reported in PC-12 cells stimulated with NGF, there is no evidence for a tyrosine kinase domain in the NGF receptor. ACTH has also been reported to stimulate the turnover of a glycosyl-PI in calf adrenal glomerulosa cells.[34] Since there are no data about the hydrolysis by PI-PlC or sensitivity to nitrous acid deamination of this glycolipid in glomerulosa cells, the possibility exists that it might be a different type of lipid. Moreover, it would be interesting to know if the effects of these various hormones on glycosyl-PI hydrolysis are additive with those of insulin. In cultured pig thyroid cells, the addition of TSH produced a rapid release of an inositol phosphate-glycan and the parallel hydrolysis of the glycosyl-PI precursor.[71] This inositol phosphate-glycan isolated from pig thyroid cells seems to be similar in its analytical and biological properties to the putative mediator of insulin action. Thus, the addition of this molecule to adipocytes decreased both glycerol release and cAMP accumulation,[71] as previously reported with the "insulin mediator".[45,48]

Transformation of rat-1 cells with membrane-bound oncogenes, mutant H-*ras* and v-*src*, but not with a nuclear oncogene, c-*myc*, resulted in a severalfold elevation of the basal levels of glycosyl-PI.[82] Moreover, analysis of different cell lines, transformed by mutant H-*ras*, shows that the observed elevation correlates with the transformed phenotype of these cells. These results suggest a role for glycosyl-PI in the transformation process or, vice versa, a role for the oncogenic protein(s) in the metabolism of this glycolipid. The mechanism(s) by which the levels of glycosyl-PI are elevated in H-*ras* and v-*src*-transformed cells remains unknown. Available data on the kinetics of glycosyl-PI synthesis and degradation do not support a mechanism in which c-H-*ras* or c-*src* directly regulate the rate of hydrolysis of the insulin-sensitive glycosyl-PI, and it has been considered that regulation of glycosyl-PI levels might be a point of "cross-talk" between membrane-bound oncogenes and the insulin signal transduction pathway. This last conclusion is interesting in view of the proposed role of glycosyl-PI in mediating the effects of insulin on cellular growth and development.

A role for the phosphoinositides has been suggested in the insulin-signaling mechanism. Thus, a small, transient increase in inositol trisphosphate has been reported in fat cells after the addition of insulin.[36] However, other investigators have reported that insulin had no effect on phosphoinositide hydrolysis[37-40] or Ca^{2+} levels[41,42] in a variety of tissues, although it might stimulate the *de novo* synthesis of diacylglycerol[43] and the phosphoinositides.[38] Moreover, tyrosine phosphorylation *in vitro* of purified bovine brain PI-PlC II by purified EGF receptor could not be observed using the cytoplasmic tyrosine kinase domain of the insulin receptor,[61,65] which further indicates that insulin signaling is not coupled to phosphatidylinositol turnover. In swiss 3T3 cells, insulin had no effect of its own on phosphatidylinositol turnover, but increased the turnover of inositol lipids due to acute bombesin stimulation.[63] The mechanism of this effect of insulin on phosphatidylinositol turnover is unknown.

A variety of studies indicate that diabetes leads to changes in enzymes involved in the metabolism of inositol phospholipids.[66] Streptozotocin-induced diabetes in the rat has been shown to lower sciatic nerve *myo*-inositol levels[67] and the levels of inositol phospholipids.[68] Moreover, a decrease in inositol phospholipids of sciatic nerve obtained postmortem from diabetic and control human subjects has also been reported.[69] Changes in enzymes of inositol and phosphoinositide metabolism have also been described in diabetes.[66] Thus, inositol synthetase activity in the testis, CDP-diacylglycerol-inositol phosphatidyltransferase, and phosphatidylinositol-4-phosphate kinase in brain and sciatic nerve from rats have also been found to be decreased.[70] The causes of these alterations in phosphoinositide metabolism in diabetes are not understood.

IV. INSULIN-LIKE EFFECTS OF INOSITOL PHOSPHO-OLIGOSACCHARIDES

The results mentioned above indicate that insulin stimulates the generation, at the outer cell surface, of an inositol phospho-oligosaccharide (POS) which is the polar headgroup of a glycosyl-PI. In 1986, Mato's laboratory reported that the addition of POS to isolated rat adipocytes had insulin-like effects on phospholipid methyltransferase,[44] the enzyme that converts phosphatidylethanolamine into phosphatidylcholine by three successive N-methylations, S-adenosyl-L-methionine being the methyl donor (see Chapter 6). Since then, this or similar preparations of POS have been shown to have insulin-like effects on lipolysis,[45] lipogenesis,[46] and pyruvate dehydrogenase[47] in intact adipocytes, and glycogen phosphorylase, pyruvate kinase, and cAMP levels[48] in intact hepatocytes. These effects of POS are time and dose dependent.[45] POS, however, has no effect on glucose transport in adipocytes.[45,46] It is interesting to note, however, that all the biological effects of POS mentioned above were carried out with a molecule isolated from rat liver and that, in this tissue, insulin has no effect on glucose transport. Therefore, it would be important to know if POS isolated from adipocytes might facilitate glucose transport in this tissue. In this regard, it is worth noticing that glucose uptake has been stimulated by the addition to rat adipocytes of an inositol-phospho-oligosaccharide fraction isolated from Actovegin, a drug derived from calf blood.[49] Moreover, diacylglycerol has been shown to act as an activator of glucose transport in adipocytes when applied outside the cell,[50] and addition of phospholipase C to intact adipocytes also stimulated glucose transport.[72] This raises the interesting possibility that insulin generates two types of biochemical signals: POS, which mediates a number of biological effects of insulin, and diacylglycerol, which mediates the effects of the hormone on glucose uptake. It has also been shown that in BC3H1 myocytes, insulin stimulates the *de novo* synthesis of phosphatidic acid and diacylglycerol.[43,51] The contribution of the diacylglycerol formed by hydrolysis of glycosyl-PI to this increase in the *de novo* synthesis of diacylglycerol is probably negligible, and the biological function of both pools must also be different. Sphingosine, a reversible inhibitor of protein kinase C which acts competitively with respect to activators such as phorbol ester and diacylglycerol (see Chapter 9), has been shown to inhibit insulin-stimulated glucose transport in 3T3-L1 fibroblasts[73] and adipocytes,[74] which is consistent with a role for diacylglycerol and protein kinase C in the activation of glucose transport by this hormone. POS has also been found to inhibit D-glucose- or 2-ketoisocapronate-induced insulin release in pancreatic islet cells from rats without affecting glucose oxidation or ^{45}Ca net uptake.[52,53] The relative extent of 2-ketoisocapronate-induced insulin release was unaffected by either the concentration of D-glucose or the presence of dibutyryl-cAMP, forskolin, or glucagon in the incubation medium. The physiological significance of this effect of POS on insulin release remains to be determined.

The effects of POS on human fibroblasts have also been investigated.[78] Human fibroblasts represent an interesting model, since the insulin has been shown to enhance the glycolytic flux in these cells through a mechanism involving the activation of a key regulatory enzyme, such as phosphofructokinase-1.[79,80] In addition, it has recently been demonstrated that the intracellular levels of fibroblast fructose-2,6-bisphosphate, a powerful activator of phosphofructokinase-1, are increased upon insulin addition and that this metabolite plays a key role in the potentiation of the glycolytic pathway brought about by the hormone.[81] Similar to insulin, POS elicited a dose-dependent stimulation of lactate output and fructose-2,6-bisphosphate content.[78] These results indicate that POS faithfully copies the effects of insulin on glucose metabolism in human fibroblasts. Moreover, these results suggest that glucose transport does not represent a rate-limiting step in the enhancement of fibroblast glucose metabolism, and that POS can act on a cell system (human fibroblasts) originated from a species phylogenetically distinct from the source of the compound (rat liver).

All the aforementioned biological effects of insulin mimicked by POS occur within seconds to minutes of the addition of the hormone. Moreover, insulin also produces a variety of long-term actions, such as the stimulation of amino acid transport, that require protein synthesis and DNA synthesis. In isolated rat hepatocytes, POS has been found to stimulate the transport of amino acids through a mechanism that requires protein synthesis, and that is additive with glucagon, indicating that this molecule also mimics some of the long-term effects of insulin.[83] It is also interesting to note that the incorporation of thymidine into DNA has been stimulated by the addition to cells of an oligosaccharide obtained from the conditioned media of Reuber hepatoma cells.[22] Whether this compound is related to POS remains to be determined. Insulin and NGF produce highly specific effects on the otic vesicle. This is a transient embryonic structure occurring during the early development of the inner ear. Insulin typically potentiates the proliferative effects of other growth factors on the epithelium of quiescent otic vesicles, and NGF behaves as a potent mitogen in the gangliolar cell population. These effects of insulin and NGF on the otic vesicle are faithfully copied by POS.[84] Since NGF has been shown to stimulate the turnover of glycosyl-PI,[33] these results raise the interesting possibility that POS might mediate the mitogenic effect of various growth factors.

Insulin controls cellular metabolism by regulating the state of serine and threonine phosphorylation of a variety of proteins and rate-limiting enzymes like glycogen phosphorylase, hormone-sensitive lipase, and ATP citrate lyase.[1,2] If insulin action is dependent on the generation of POS, this molecule(s) might also mimic insulin's effect on serine and threonine phosphorylation of key target enzymes and proteins. Insulin stimulates the phosphorylation of a protein with an M_r of 116 kDa in intact adipocytes identified as ATP citrate lyase.[54] POS mimics this effect of insulin when added to intact cells, an effect which is dose and time dependent, being half-maximal at about 4 μM and after 15 min.[55] Insulin also reduces the phosphorylation of an 84-kDa phosphoprotein identified as hormone-sensitive lipase.[56] POS also reproduces this insulin dephosphorylation effect in a dose- and time-dependent process, which is half-maximal at about 4 μM and after 5 min.[55] In the case of ATP citrate lyase, the site of phosphorylation stimulated by insulin and POS was also studied and found to be the same.[57] Isoproterenol, via the elevation of the intracellular concentration of cAMP, increases the phosphorylation of a specific subset of phosphoproteins, and insulin blocks this isoproterenol-induced increase in phosphorylation of the phosphoproteins with M_r of 97 kDa (glycogen phosphorylase), 84 kDa (hormone-sensitive lipase), 65 kDa and 47 kDa. This effect of insulin is also reproduced by POS, with the half-maximal effect being at about 3 μM.[55,58] These results indicate that in adipocytes POS faithfully copies the insulin-directed effects on the phosphorylation/dephosphorylation of target proteins of the hormone. The conclusion of these results is also that POS stimulates the same signaling pathways as does insulin. Since both cAMP-dependent and -independent effects are mimicked by POS, the hydrolytic production of POS might be an early step in the insulin-signaling mechanism. In favor of this hypothesis is the observation that pretreatment of isolated rat hepatocytes with β-galactosidase, which has been mentioned above, removed the majority of the glycosyl-PI in these cells and blocked the effect of insulin on glucagon-inhibited pyruvate kinase activity, although it had no effect on the inhibition by glucagon of the enzyme.[18] To confirm the hypothesis that POS mediates some of the biological effects of insulin, it is imperative to obtain the complete structure of this molecule(s) and to have available synthetic oligosaccharides to test for their biological activity. In this respect, it has been recently synthesized the pentasaccharide core of the glycosyl-PI anchor found in *T. brucei*.[75] Whether this or related molecules have biological activity is not yet known.

V. MECHANISM OF ACTION OF POS: INTRACELLULAR VS. EXTRACELLULAR ACTION

Since POS has insulin-like effects when added to intact cells, the question is whether POS acts intracellularly or through its interaction with a cell surface receptor. Evidence in favor of an intracellular action comes from the observation that POS affects the activity of several cellular enzymes, such as cyclic nucleotide phosphodiesterase, pyruvate dehydrogenase, adenylate cyclase,[7,14,47,58,59] cAMP-dependent protein kinase,[60] and casein kinase II,[85] in a manner similar to insulin when added to intact cells. In the case of the cAMP-dependent protein kinase, the inhibition of the activity was dose dependent, with the half-maximal effect at 2 μM.[60] This effect was demonstrated in the presence of 1 mM ATP, indicating that POS acts at physiological concentrations of ATP and that it does not compete with this nucleotide for the same binding site in the kinase. In the case of casein kinase II, whereas at 2 μM POS stimulated the enzyme, higher concentrations of this molecule were inhibitory, 50% inhibition of the enzyme was obtained at 15 μM POS. This biphasic effect of POS on casein kinase II was observed using as substrate casein or the specific peptide for casein kinase II, Arg-Arg-Arg-Glu-Glu-Glu-Thr-Glu-Glu-Glu. Modulation of casein kinase II activity by POS was specific and was not observed with other compounds structurally related to POS. The effect of POS on casein kinase II was still observed after gel filtration, a situation similar to that found in 3T3-L1 cells and in BALB/c 3T3 fibroblasts after insulin addition,[76,77] which supports the concept that these signals modulate casein kinase II through the generation of POS. POS, however, has no effect on rat brain protein kinase C[50] or on insulin-receptor autophosphorylation,[22] indicating that these effects on protein phosphorylation are specific.

The uptake of POS has been examined in intact rat hepatocytes using [^{3}H]galactose-phospho-oligosaccharide.[86] The labeled molecule was prepared by reacting the glycosyl phosphatidylinositol precursor purified from rat liver with NaB[^{3}H]$_4$ and galactose oxidase, followed by treatment of the radioactive glycolipid with *Bacillus thuringensis* phosphatidylinositol-specific phospholipase C. [^{3}H]galactose-phospho-oligosaccharide was taken up by the isolated hepatocytes at 37°C, but not at 4°C. The uptake of this molecule was concentration and time dependent, and reached a maximum within 5 to 10 min of the addition of the labeled phospho-oligosaccharide. Pretreatment of the hepatocytes for 5 min with insulin increased approximately two-fold the uptake of the phospho-oligosaccharide. The concentration of phospho-oligosaccharide required for half-maximal uptake was about 200 μM in both the absence and presence of insulin. The effect of insulin on phospho-oligosaccharide uptake was dose dependent, with the half-maximal effect at about 1nM. The uptake of the phospho-oligosaccharide (10 μM final concentration) was specific and could not be blocked by the addition to the incubation media of a tenfold excess of *myo*-inositol, *myo*-inositol-1-phosphate, or *myo*-inositol-2-phosphate. Mannose, *myo*-inositol hexakis-phosphate, and glucosamine (100 μM final concentration) inhibited phospho-oligosaccharide uptake (10 μM final concentration) by only 34, 43, and 55%, respectively. These results suggest that the generation of this phospho-oligosaccharide at the cell surface and its transport to the cell interior might be an important step in the insulin-signaling mechanism.

In conclusion, the present evidence supports a model where insulin promotes the hydrolysis of a glycosyl-PI at the outer surface of the cell, with release of its polar headgroup and diacylglycerol (Figure 2). The inositol phospho-oligosaccharide that forms the polar headgroup enters the cell and modulates the state of phosphorylation of key target proteins and mediates a number of short- and long-term effects of the hormone.

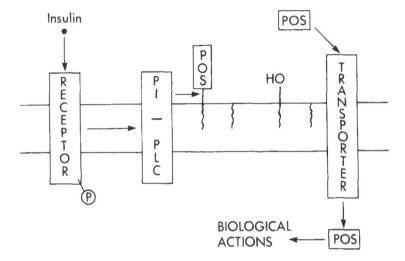

FIGURE 2.   Model for insulin action where the hormone promotes the hydrolysis of a glycosyl-PI at the outer surface of the cell, with release of its polar headgroup and diacylglycerol. The inositol phospho-oligosaccharide that forms the polar headgroup enters the cell and modulates the state of phosphorylation of key target proteins and mediates a number of short- and long-term effects of the hormone. The molecule of diacylglycerol generated at the outer surface of the plasma membrane might be involved in the control of glucose transport.

REFERENCES

1. **Denton, R. M.,** Early events in insulin actions, in *Advances in Cyclic Nucleotide Research,* Vol. 20, Greengard, P. and Robison, G. A., Eds., Raven Press, New York, 1986, 293.
2. **Rosen, O. M.,** After insulin binds, *Science,* 237, 1452, 1987.
3. **Petruzzelli, L., Herrera, R., and Rosen, O. M.,** Insulin receptor is an insulin-dependent tyrosine kinase: copurification of insulin-binding activity, and protein kinase activity to homogeneity from human placenta, *Proc. Natl. Acad. Sci. U.S.A.,* 81, 3327, 1981.
4. **Chou, C. K., Dull, T. J., Rusell, D. S., Gherzi, R., Lebwohl, D., Ullrich, A., and Rosen, O. M.,** Human insulin receptors mutated at the ATP-binding site lack protein tyrosine kinase activity and fail to mediate postreceptor effects of insulin, *J. Biol. Chem.,* 262, 1842, 1987.
5. **Russell, D. S., Gherzi, R., Johnson, E. L., Chou, C. K., and Rosen, O.,** The protein-tyrosine kinase activity of the insulin receptor is necessary for insulin-mediated receptor down-regulation, *J. Biol. Chem.,* 262, 11833, 1987.
6. **Mato, J. M.,** Insulin mediators revisited, *Cell. Signal.,* 1, 143, 1989.
7. **Saltiel, A. R., Fox, J. A., Sherline, P., and Cuatrecasas, P.,** Insulin stimulates the hydrolysis of a novel membrane glycolipid causing the generation of cAMP phosphodiesterase modulators, *Science,* 233, 967, 1986.
8. **Mato, J. M., Kelly, K. L., Abler, A., and Jarett, L.,** Identification of a novel insulin-sensitive glycophospholipid from H35 hepatoma cells, *J. Biol. Chem.,* 262, 2131, 1987.
9. **Gaulton, G., Kelly, K. L., Mato, J. M., and Jarett, L.,** Regulation and function of an insulin-sensitive glycosyl-phosphatidylinositol during T lymphocyte activation, *Cell,* 53, 963, 1988.
10. **Villalba, M., Alvarez, J. F., Russell, D., Mato, J. M., and Rosen, O.,** Hydrolysis of glycosyl-phosphatidylinositol in response to insulin is reduced in cells bearing kinase-deficient insulin receptors, *Growth Factors,* 2, 91, 1990.
11. **Cross, G. A. M.,** Eukaryotic protein modification and membrane attachment via phosphatidylinositol, *Cell,* 48, 179, 1987.
12. **Ferguson, M. A. J. and Williams, A. F.,** Cell-surface anchoring of proteins via glycosyl-phosphatidylinositol structures, *Annu. Rev. Biochem.,* 57, 285, 1988.
13. **Mato, J. M., Kelly, K. L., Abler, A., Jarett, L., Corkey, B. E., Cashel, J. A., and Zopf, D.,** Partial structure of an insulin-sensitive glycophospholipid, *Biochem. Biophys. Res. Commun.,* 146, 764, 1987.

14. **Larner, J., Huang, L. C., Schwartz, C. F. W., Oswald, A. S., Shen, T. Y., Kinter, M., Tang, G., and Zeller, K.**, Rat liver insulin mediator which stimulates pyruvate dehydrogenase phosphatase contains galactosamine and D-chiroinositol, *Biochem. Biophys. Res. Commun.*, 151, 1416, 1988.

15. **Merida, I., Corrales, F. J., Clemente, R., Ruiz-Albusac, J. M., Villalba, M., and Mato, J. M.**, Different phosphorylated forms of an insulin-sensitive glycosylphosphatidylinositol from rat hepatocytes, *FEBS Lett.*, 236, 251, 1988.

16. **Futerman, A. H., Low, M. G., Ackermann, K. E., Sherman, W. R., and Silman, I.**, Biochemical behaviour and structural characteristics of membrane-bound acetylcholinesterase from *Torpedo* electric organ, *Biochem. Biophys. Res. Commun.*, 129, 312, 1985.

17. **Low, M. G., Futerman, A. H., Ackerman, K. E., Sherman, W. R., and Silman, I.**, Removal of covalently bound inositol from *Torpedo* acetylcholinesterase and mammalian alkaline phosphatase by deamination with nitrous acid, *Biochem. J.*, 241, 615, 1987.

18. **Varela, I., Alvarez, J. F., Ruiz-Albusac, J. M., and Mato, J. M.**, Asymmetric distribution of the phosphatidylinositol-linked phospho-oligosaccharide that mimics insulin action in the plasma membrane, *Eur. J. Biochem.*, 188, 213, 1990.

19. **McConville, M. and Bacic, A.**, A family of glycoinositol phospholipids from *Leishmania major*. Isolation, characterization and antigenicity, *J. Biol. Chem.*, 264, 757, 1989.

20. **Alvarez, J. F., Varela, I., Ruiz-Albusac, J. M., and Mato, J. M.**, Localisation of the insulin-sensitive phosphatidylinositol glycan at the outer surface of the cell membrane, *Biochem. Biophys. Res. Commun.*, 152, 1455, 1988.

21. **Romero, G., Lutrell, L., Rogol, A., Zeller, K., Hewlett, E., and Larner, J.**, Phosphatidylinositol-glycan anchors of membrane proteins: potential precursors of insulin mediators, *Science*, 240, 509, 1988.

22. **Witters, L. A. and Watts, T. D.**, An autocrine factor from reuber hepatoma cells that stimulates DNA synthesis and acetyl-CoA carboxylase, *J. Biol. Chem.*, 263, 8027, 1988.

23. **Krakow, J. L., Hereld, D., Bangs, J. D., Hart, G. W., and Englund, P. T.**, Identification of a glycolipid precursor of the *Trypanosoma brucei* variant surface glycoprotein, *J. Biol. Chem.*, 261, 12147, 1986.

24. **Menon, A. K., Mayor, S., Ferguson, M. A. J., Duszenko, M., and Cross, G. A. M.**, Candidate glycolipid precursor for the glycosyl-phosphatidylinositol membrane anchor of *Trypanosoma brucei* variant surface glycoprotein, *J. Biol. Chem.*, 263, 1970, 1988.

25. **McNealy, T. B., Rosen, G., Londner, M. V., and Turco, S. J.**, Inhibitory effects on protein kinase C activity by lipophosphoglycan fragments and glycophosphatidylinositol antigens of the protozoan parasite *Leishmania*, *Biochem. J.*, 259, 601, 1989.

26. **Chan, B. L., Lisanti, M. P., Rodriguez-Boulan, E., and Saltiel, A. R.**, Insulin-stimulated release of lipoprotein lipase by metabolism of its phosphatidylinositol anchor, *Science*, 241, 1670, 1988.

27. **Fox, J. A., Soliz, N. M., and Saltiel, A. R.**, Purification of a phosphatidylinositol-glycan-specific phospholipase C from liver plasma membranes: a possible target of insulin action, *Proc. Natl. Acad. Sci. U.S.A.*, 84, 2663, 1987.

28. **Wahl, M. I., Daniel, T. O., and Carpenter, G.**, Antiphosphotyrosine recovery of phospholipase C activity after EGF treatment of A-431 cells, *Science*, 241, 968, 1988.

29. **Wahl, M. I., Nishibe, S., Suh, P. G., Rhee, S. G., and Carpenter, G.**, Epidermal growth factor stimulates tyrosine phosphorylation of phospholipase C-II independently of internalization and extracellular Ca^{2+}, *Proc. Natl. Acad. Sci. U.S.A.*, 86, 1568, 1989.

30. **Ruiz-Albusac, J. M., Zueco, J. A., Velazquez, E., Alvarez, J. F., Mato, J. M., and Blazquez, E.**, Insulin induces a similar reduction in the concentrations of its own receptor and of an insulin-sensitive glycosyl-phosphatidylinositol in isolated rat hepatocytes, *FEBS Lett.*, 258, 281, 1989.

31. **Cabello, M. A., Sanchez-Arias, J. A., Liras, A., Mato, J. M., and Feliu, J. E.**, Effect of adrenalectomy and glucocorticoid treatment on the levels of an insulin-sensitive glycosyl-phosphatidylinositol in isolated rat hepatocytes, *Cel. Mol. Endocrinol.*, 68, R1, 1990.

32. **Farese, R. V., Nair, G. P., Stadaert, M. L., and Cooper, D. R.**, Epidermal growth factor and insulin-like growth factor-I stimulate the hydrolysis of the insulin-sensitive phosphatidylinositol-glycan in BC3H1 myocytes, *Biochem. Biophys. Res. Commun.*, 156, 1346, 1988.

33. **Chan, B. L., Chao, M. V., and Saltiel, A. R.**, Nerve growth factor stimulates the hydrolysis of glycosyl-phosphatidylinositol in PC-12 cells: a mechanism of protein kinase C regulation, *Proc. Natl. Acad. Sci. U.S.A.*, 86, 1756, 1989.

34. **Cozza, E. N., Vila, M. C., Gomez-Sanchez, C. E., and Farese, R. V.**, ACTH stimulates turnover of the phosphatidylinositol-glycan, *Biochem. Biophys. Res. Commun.*, 157, 585, 1988.

35. **Koepfer-Hobelsberger, B. and Wieland, O. H.**, Insulin activates phospholipase C in fat cells: similarity with the activation of pyruvate dehydrogenase, *Mol. Cell. Endocrinol.*, 36, 123, 1984.

36. **Farese, R. V., Kuo, J. Y., Babischkin, J. S., and Davis, J. S.**, Insulin provokes a transient activation of phospholipase C in the rat epididymal fat pad, *J. Biol. Chem.*, 261, 8589, 1986.

37. **Creba, J. A., Downs, C. P., Howkins, P. T., Brewster, G., Michell, R. H., and Kirk, C. J.**, Rapid breakdown of phosphatidylinositol 4-phosphate and phosphatidylinositol 4,5-bisphosphate in rat hepatocytes stimulated by vasopressin and other Ca^{2+}-mobilizing hormones, *Biochem. J.*, 212, 733, 1983.

38. **Pennington, S. R. and Martin, B. R.**, Insulin-stimulated phosphoinositide metabolism in isolated fat cells, *J. Biol. Chem.*, 260, 11039, 1985.

39. **Besterman, J. M., Watson, S. P., and Cuatrecasas, P.**, Lack of association of EGF-, insulin- and serum-induced mitogenesis with stimulation of phosphoinositide degradation in BALB/c 3T3 fibroblasts, *J. Biol. Chem.*, 261, 723, 1986.

40. **Thaker, J. K., DiMarchi, R., MacDonald, K., and Caro, J. F.**, Effect of insulin and insulin-like growth factors I and II on phosphatidylinositol and phosphatidylinositol 4,5-bisphosphate breakdown in liver from humans with and without type II diabetes, *J. Biol. Chem.*, 264, 7169, 1989.

41. **Moolenar, W. H., Tertoolen, L. G. J., and de Laat, S. W.**, Growth factors immediately raise cytoplasmic free Ca^{2+} in human fibroblasts, *J. Biol. Chem.*, 259, 8066, 1984.

42. **Thomas, A. P., Martin-Reguero, A., and Williamson, J. R.**, Interactions between insulin and α_1-adrenergic agents in the regulation of glycogen metabolism in isolated hepatocytes, *J. Biol. Chem.*, 260, 5963, 1985.

43. **Farese, R. V., Barnes, D. E., Davis, J. S., Standaert, M. L., and Pollet, R. J.**, Effects of insulin and protein synthesis inhibitors on phospholipid metabolism, diacylglycerol levels, and pyruvate dehydrogenase activity in BC3H-1 cultured myocytes, *J. Biol. Chem.*, 259, 7094, 1984.

44. **Kelly, K. L., Mato, J. M., and Jarett, L.**, The polar head group of a novel insulin-sensitive glyco-phospholipid mimics insulin action on phospholipid methyltransferase, *FEBS Lett.*, 209, 238, 1986.

45. **Kelly, K. L., Mato, J. M., Merida, I., and Jarett, L.**, Glucose transport and antilipolysis are differentially regulated by the polar head group of an insulin-sensitive glycophospholipid, *Proc. Natl. Acad. Sci. U.S.A.*, 84, 6404, 1987.

46. **Saltiel, A. R. and Sorbara-Cazan, L. R.**, Inositol glycans mimic the action of insulin on glucose utilization in rat adipocytes, *Biochem. Biophys. Res. Commun.*, 149, 1084, 1987.

47. **Gottschalk, K. W. and Jarett, L.**, The insulinomimetic effects of the polar head group of an insulin-sensitive glycophospholipid on pyruvate dehydrogenase in both subcellular and whole cell assays, *Arch. Biochem. Biophys.*, 261, 175, 1988.

48. **Alvarez, J. F., Cabello, M. A., Feliu, J. E., and Mato, J. M.**, A phospho-oligosaccharide mimics insulin action on glycogen phosphorylase and pyruvate kinase activities in isolated rat hepatocytes, *Biochem. Biophys. Res. Commun.*, 147, 765, 1987.

49. **Machicao, F., Mushack, J., Seffer, E., Ermel, B., and Haring, H. V.**, Mannose, glucosamine and inositol monophosphate inhibit the effects of insulin on lipogenesis: Further evidence for a role of inositol phosphate-oligosaccharides in insulin action, *Biochem. J.*, 266, 909, 1990.

50. **Strälfors, P.**, Insulin stimulation of glucose uptake can be mediated by diacylglycerol in adipocytes, *Nature*, 335, 554, 1988.

51. **Farese, R. V., Cooper, D. R., Konda, T. S., Nair, G., Standaert, M. L., Davis, J. S., and Pollet, R. J.**, Mechanisms whereby insulin increases diacylglycerol in BC3H-1 myocytes, *Biochem. J.*, 256, 175, 1988.

52. **Albor, A., Camara, J., Valverde, I., Mato, J. M., and Malaisse, W. J.**, Inhibition of insulin release by a putative insulin-mediator in pancreatic islet cells, *Med. Sci. Res.*, 17, 161, 1989.

53. **Malaisse, W. J., Albor, A., Valverde, I., Sener, A., and Mato, J. M.**, Effect of a phospho-oligosaccharide putative insulin messenger on insulin release in rats, *Diabetologia*, 32, 295, 1989.

54. **Ramakrishna, S. and Benjamin, W. B.**, Fat cell protein phosphorylation. Identification of phosphoprotein-2 as ATP-citrate lyase, *J. Biol. Chem.*, 254, 9232, 1979.

55. **Alemany, S., Mato, J. M., and Strälfors, P.**, Phospho-dephospho-control by insulin is mimicked by a phospho-oligosaccharide in adipocytes, *Nature*, 330, 77, 1987.

56. **Strälfors, P., Björgell, P., and Belfrage, P.**, Hormonal regulation of hormone-sensitive lipase in intact adipocytes: identification of phosphorylation sites and effects on the phosphorylation by lipolytic hormones and insulin, *Proc. Natl. Acad. Sci. U.S.A.*, 81, 3317, 1984.

57. **Puerta, J., Alemany, S., and Mato, J. M.**, The inositol phospho-oligosaccharide that forms the polar head group of the insulin-sensitive glycosyl-PI modulates the state of phosphorylation of key target proteins of the hormone, *Adv. Enzyme Regul.*, 30, 109, 1990.

58. **Kelly, K. L., Merida, I., Wong, E. H. A., DiCenzo, D., and Mato, J. M.**, A phospho-oligosaccharide mimics the effect of insulin to inhibit isoproterenol-dependent phosphorylation of phospholipid methyltransferase in isolated adipocytes, *J. Biol. Chem.*, 262, 15282, 1987.

58. **Saltiel, A. R. and Cuatrecasas, P.**, Insulin stimulates the generation from hepatic membranes of modulators derived from an inositol glycolipid, *Proc. Natl. Acad. Sci. U.S.A.*, 83, 5793, 1986.

59. **Saltiel, A. R.**, Insulin generates an enzyme modulator from hepatic plasma membranes: regulation of adenosine 3',5'-monophosphate phosphodiesterase, pyruvate dehydrogenase, and adenylate cyclase, *Endocrinology*, 120, 967, 1987.

60. **Villalba, M., Kelly, K. L., and Mato, J. M.**, Inhibition of cyclic AMP-dependent protein kinase by the polar head group of an insulin-sensitive glycophospholipid, *Biochim. Biophys. Acta*, 968, 69, 1988.

61. **Nishibe, S., Wahl, M. I., Rhee, S. G., and Carpenter, G.,** Tyrosine phosphorylation of phospholipase C-II *in vitro* by the epidermal growth factor receptor, *J. Biol. Chem.*, 264, 10335, 1989.

62. **Bülow, R., Griffiths, G., Webster, P., Stierhof, Y. D., Opperdoes, F. R., and Overath, P.,** Intracellular localization of the glycosyl-phosphatidylinositol-specific phospholipase C of *Trypanosome brucei, J. Cell. Sci.*, 93, 233, 1989.

63. **Heslop, J. P., Blakely, D. M., Brown, K. D., Irvine, R. F., and Berridge, M. J.,** Effects of bombesin and insulin on inositol (1,4,5)trisphosphate and inositol (1,3,4)trisphosphate formation in swiss 3T3 cells, *Cell,* 47, 703, 1986.

64. **Nishibe, S., Wahl, M. I., Rhee, S. G., and Carpenter, G.,** Tyrosine phosphorylation of phospholipase C-II *in vitro* by the epidermal growth factor receptor, *J. Biol. Chem.*, 264, 10335, 1989.

65. **Meisenhelder, J., Suh, P. G., Rhee, S. G., and Hunter, T.,** Phospholipase C-τ is a substrate for the PDGF and EGF receptor protein-tyrosine kinases in vivo and in vitro, *Cell,* 57, 1109, 1989.

66. **Sherman, W. R.,** Inositol homeostasis, lithium and diabetes, in *Inositol Lipids in Cell Signalling,* Michell, R. H., Drummond, A. H., and Downes, C. P., Eds., Academic Press, London, 1989, 39.

67. **Greene, D. A., De Jesus, P. V., and Winegrad, A. I.,** Effects of insulin and dietary myoinositol on impaired peripheral motor nerve conduction velocity in acute streptozotocin diabetes, *J. Clin. Invest.*, 55, 1326, 1975.

68. **Palmano, K. P., Whiting, P. H., and Hawthorne, J. N.,** Free and lipid myo-inositol in tissues from rats with acute and less severe streptozotocin-induced diabetes, *Biochem. J.,* 167, 229, 1977.

69. **Mayhew, J. A., Gillon, K. R. W., and Hawthorne, J. N.,** Free and lipid inositol, sorbitol and sugars in sciatic nerve obtained post-mortem from diabetic patients and control subjects, *Diabetologia,* 24, 13, 1983.

70. **Whiting, P. H., Palmano, K. P., and Hawthorne, J. N.,** Enzymes of myo-inositol lipid metabolism in rats with streptozotocin-induced diabetes, *Biochem. J.,* 179, 549, 1979.

71. **Martiny, L., Antonicelli, F., Thuilliez, B., Lambert, P., Jacquemin, C., and Haye, B.,** Control by TSH of the production by thyroid cells of an inositol phosphate-glycan, *Cell. Signal.,* 2, 21, 1990.

72. **Obermaier-Kusser, B., Mühlbacher, C., Mushack, J., Rattenhuber, E., Fehlmann, M., Haring, H. U.,** Regulation of glucose carrier activity by $AlCl_3$ and phospholipase C in fat-cells, *Biochem. J.,* 256, 515, 1988.

73. **Nelson, D. and Murray, D.,** Sphingolipids inhibit insulin and phorbol ester stimulated uptake of 2-deoxyglucose, *Biochem. Biophys. Res. Commun.,* 138, 463, 1986.

74. **Robertson, D. G., DiGirolano, M., Merrill, A. H., and Lambeth, J. D.,** Insulin-stimulated hexose transport and glucose oxidation in rat adipocytes is inhibited by sphingosine at a step after insulin binding, *J. Biol. Chem.,* 264, 6778, 1989.

75. **Mootoo, D. R., Konradsson, P., and Fraser-Reid, B.,** *N*-pentyl glycosides facilitate a stereoselective synthesis of the pentasaccharide core of the protein membrane anchor in *Trypanosoma brucei, J. Am. Chem. Soc.,* 111, 8540, 1989.

76. **Sommercorn, J. and Krebs, E. G.,** Classification of protein kinase into messenger-dependent and in-dependent kinases. The regulation of independent kinases, *Adv. Exp. Med. Biol.,* 231, 403, 1988.

77. **Klarlund, J. K. and Czech, M. P.,** Insulin-like growth factor I and insulin rapidly increase casein kinase II activity in BALB/c 3T3 fibroblasts, *J. Biol. Chem.,* 263, 15872, 1988.

78. **Bruni, P., Meacci, E., Avila, M., Vasta, V., Farnararo, M., Mato, J. M., and Varela, I.,** A phospho-oligosaccharide can reproduce the stimulatory effect of insulin on glycolytic flux in human fibroblasts, *Biochem. Biophys. Res. Commun.,* 166, 765, 1990.

79. **Diamond, I., Legg, A., Schneider, J. A., and Rozengurt, E.,** Glycolysis in quiescent cultures of 3T3 cells. Stimulation by serum, epidermal growth factor and insulin in intact cells and resistence of the stimulation after cell homogenization, *J. Biol. Chem.,* 253, 866, 1978.

80. **Schneider, J. A., Diamond, I., and Rozengurt, E.,** Glycolysis in quiescent cultures of 3T3 cells. Addition of serum, EGF and insulin increases the activity of phosphofructokinase in a protein synthesis-independent manner, *J. Biol. Chem.,* 253, 872, 1978.

81. **Bosca, L., Rousseau, G. G., and Hue, L.,** Phorbol 12-myristate 13-acetate and insulin increase the concentration of fructose, 2,6 bisphosphate and stimulate glycolysis in chicken embryo fibroblasts, *Proc. Natl. Acad. Sci. U.S.A.,* 82, 6440, 1985.

82. **Burgering, B. M. T., Alemany, S., Clemente, R., Bos, J. L., and Mato, J. M.,** unpublished results.

83. **Varela, I., Avila, M., Mato, J. M., and Hue, L.,** Insulin-induced phospho-oligosaccharide stimulates amino acid transport in isolated rat hepatocytes, *Biochem. J.,* 267, 541, 1990.

84. **Varela, I., Avila, M., Miner, C., Represa, J., Mato, J. M., and Giraldez, F.,** unpublished results.

85. **Alemany, S., Puerta, J., Guadaño, A., and Mato, J. M.,** Modulation of casein kinase II activity by the polar head group of an insulin sensitive glycosyl-phosphatidylinositol, *J. Biol. Chem.,* 265, 4849, 1990.

86. **Alvarez, J. F. and Mato, J. M.,** unpublished results.

Chapter 9

SPHINGOLIPIDS IN CELLULAR SIGNALING

José M. Mato

TABLE OF CONTENTS

I. INTRODUCTION

The simplest class of sphingolipids are sphingomyelins, a major component of animal cell membranes and serum lipoproteins. The structure of sphingomyelin, *N*-acylsphingosine-1-phosphorylcholine or ceramide-1-phosphorylcholine, is shown in Figure 1.[1,2] Like the phosphatidylcholine molecules, sphingomyelins consist of two hydrophobic groups (the sphingosine group and the acyl group) and a phosphorylcholine group. Sphingomyelins occurring in biological membranes differ in the nature of the sphingosine base and the acyl group. The most common sphingosine is the 18-carbon aminediol, 1,3-dihydroxy-2-amino-4-octodecene. This molecule has a *trans* double bond between carbons 4 and 5. The dihydro derivative of this base, 1,3-dihydroxy-2-amino octadecane, is also present in biological membranes in small amounts, and is known by the name of sphinganine.[3-5] Phytosphingosine (1,3,4-hydroxy-2-aminooctadecane) has also been detected in bovine kidney.[6] These molecules are collectively referred to as long-chain sphingoid bases. The fatty acids commonly found in sphingomyelin are palmitic (16:0), nervonoyl (24:1), 22:0, and 24:0.[7] In brain, the most common fatty acid found is stearic acid (18:0), nervonoyl (24:1) and 24:0 being less common.[8] An interesting difference in the fatty acid composition of phosphoglycerolipids is the presence of longer fatty acid molecules (22:0, 24:0, and 24:1) in sphingomyelins. Little is known about the factors that control the alkyl chain length and the degree of unsaturation of the sphingolipids. Differences in the long-chain sphingoid base composition of sphingomyelin have been observed in rats bearing Morris hepatoma 7777, which suggests the existence of variations in the alkyl composition of sphingolipids in neoplasia.[9]

Except for sphingomyelin, which has a phosphorylcholine as the polar headgroup, and ceramide, which has a hydroxyl group as the polar headgroup, all other sphingolycolipids, which are named glycolipids, contain polar headgroups formed by carbohydrates. Glycolipids are a group of ubiquitous membrane lipids formed from an *N*-acyl sphingosine — a ceramide glycosidically bound to a single hexose or a complex oligosaccharide molecule through the hydroxyl group at the C-1 position (Figure 1). The sphingosine and acyl groups form the hydrophobic core of the molecule and the hexoses or the oligosaccharide moiety contributes to the hydrophillic portion. At least 300 different glycolipids exist in various mammalian cell types. Five subclasses of glycolipids can be differentiated, with basically different oligosaccharide moieties. These are named ganglio-, globo-, lacto-, gala-, and mucoglycoceramides. Within each subclass, variations in the oligosaccharide chain give rise to a large variety of glycolipids, particularly in the lacto series. The oligosaccharide structure of several glycolipids is shown in Table 4 of Chapter 2. All gangliosides have a negative charge at physiological pH due to the presence of sialic acid residues. Detailed descriptions of the glycolipids can be found in several reviews.[10,11]

The majority of the glycolipids are present at the outer leaflet of the plasma membrane bilayer (see Chapter 4). This conclusion is based mainly on experiments where intact cells are labeled with galactose oxidase and NaB[^{3}H]$_4$ and the amount of label compared with that obtained with broken cell preparations.[12,13] This method, however, does not quantitatively label all the glycolipids present on the cell surface. In fact, a large proportion of the neutral glycolipids with short oligosaccharide chains are hidden at the plasma membrane, probably due to its interaction with other membrane components.[14] In human neutrophils, lactosylceramide has been shown to be predominantly associated with a granule-rich fraction,[15] which indicates that further studies are necessary on the localization of specific glycolipids. The dynamic behavior of glycolipids in biological membranes is not known in detail. The structure of the carbohydrate polar headgroup probably determines the mobility of each glycolipid. For a given glycolipid, increasing the concentration of the same species decreases its mobility, probably by interaction between the polar headgroups and aggregation.[16]

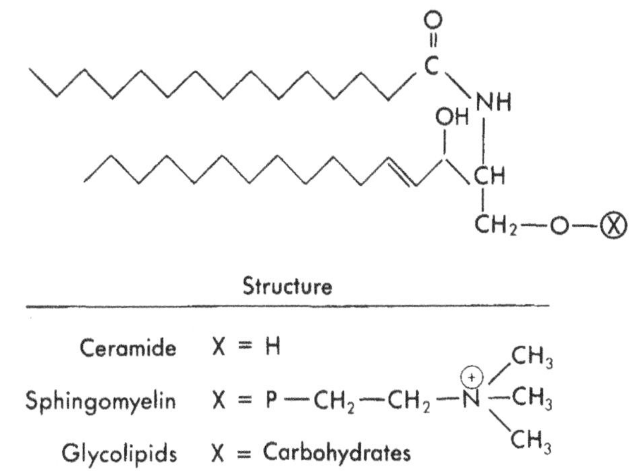

Structure

Ceramide	X = H
Sphingomyelin	X = P—CH₂—CH₂—N⁺(CH₃)₃
Glycolipids	X = Carbohydrates

FIGURE 1. Structure of sphingolipids.

II. BIOLOGICAL FUNCTIONS OF GLYCOLIPIDS

The glycolipid composition of the plasma membrane changes drastically during cell growth, cellular interaction, and transformation.[14,17] The biological function of these changes in glycolipid composition is probably related to its role in maintaining cell rigidity, to the specific interaction with surrounding membrane proteins, either directly or after suitable modification, and to the generation of cellular modulators. Three types of cellular modulators might be generated by the glycolipids: the carbohydrate moiety forming the polar headgroup, the ceramide moiety, and the lyso form of the glycolipid.

Glycolipids have been implicated in cellular growth by showing dramatic changes in ganglioside composition and biosynthesis during the cell cycle, cell differentiation, and transformation,[10,18-21] and by the observation that the addition of exogenous gangliosides (GM₃) reduces EGF-dependent cell growth.[22] It has also been reported that small modifications in the structure of GM₃ has opposite effects on cell growth and EGF-receptor tyrosine kinase activity,[23] and that the addition of certain modulators of cell growth, such as retinoids, phorbol esters, and butyrate, induce changes in ganglioside synthesis in the absence of cell growth.[24-26] Moreover, the incorporation of gangliosides and neutral glycolipids into cell membranes, by exogenous addition of the lipids, has been shown to inhibit cell growth by prolonging the G1 phase of the cell cycle.[27,28] Similarly, cell growth inhibition has been observed in BHK cells and mouse 3T3 cells treated with GM₁ or GM₃,[22,29] and a monoclonal antibody against surface galactolipids has been shown to inhibit reversibly oligodendrocyte progenitor differentiation.[76] When isolated oligodendrocyte progenitors or mixed primary cultures are grown in the presence of the antibody against surface galactolipids, myelinogenic development is blocked in a dose-dependent manner, and upon removal of the antibody, the cells rapidly resume differentiation.[76] More recently, the importance of gangliosides in cell growth has become apparent by showing that the direct interaction of the β-subunit of cholera toxin with the endogenous ganglioside GM₁ modulates cell growth in a variety of mammalian cells,[30-33] and by the observation that the interaction of GM₁ with the β-subunit of cholera toxin inhibits protein kinase C-dependent cellular proliferation.[79] Similarly, the pig edema disease toxin has been shown to bind specifically to globotetraosylceramide (GalNAcβ1-3Galα1-4Galβ1-4GlcCer).[75] In contrast to the results with GM₃, non-N-acetylated GM₃, a novel minority membrane glycolipid, has been found to be a stimulator of cell growth in A431 cells.[23,34] These results suggest the interesting possibility that removal of the N-acetyl group of GM₃ might be a mechanism to control the biological effects of this ganglioside *in vivo*.

In the brain, where ganglioside GM_1 is present in large quantities,[35] *in vivo* and *in vitro* studies have shown that gangliosides can produce axonal sprouting in the periphery as well as in the central nervous system, where they increase the survival of lesioned nigral dopamine neurons and favor recovery of the dopaminergic synaptic function in the striatum of the rat.[36,37] Moreover, GM_1 has been reported to modulate the number of neurotransmitter receptors,[38,39] and to have a potentiating effect on β-nerve growth factor-mediated responses in central cholinergic neurons,[40] which suggests that some of the trophic effects of this ganglioside might be mediated through endogenous trophic factors. Finally, in neuroblastoma cells, ganglioside GO_{1b} has been reported to have nerve-growth factor activity,[41] and the addition of GM_1 to C6 glioma cells, which are GM_1 ganglioside deficient, has been shown to inhibit DNA synthesis induced by the β-subunit of cholera toxin.[80]

The mechanism by which gangliosides modulate cell growth is not known in detail. The mitogenic effect initiated by the interaction of the β-subunit of cholera toxin with GM_1 seems to be independent of the activation of adenylate cyclase, phospholipase C, or protein kinase C, but to be mediated by an increase in intracellular Ca^{2+}, resulting from an influx from extracellular sources.[32,42] Moreover, this signaling pathway might involve a GTP-binding protein.[43] In mouse 3T3 cells, growth has been inhibited by the addition of either GM_1 or GM_3.[22] In addition, these gangliosides were found to inhibit the protein tyrosine kinase activity associated with the PDGF receptor.[22] Similarly, protein tyrosine kinase activity of EGF-receptors in A431 cells has been found to be inhibited by the addition of GM_3, but not by other gangliosides, such as GM_1, or neutral glycolipids.[44] This inhibition of the receptor tyrosine kinase activity was also demonstrated with isolated membranes and with a partially purified receptor preparation, and it has been proposed that GM_3 might bind to the EGF receptor in a manner that inhibits receptor-receptor interactions.[44] A1S, a clone of A431 cells showing a small EGF effect on cell growth together with a low degree of inhibition of EGF-receptor tyrosine kinase activity by GM_3, has been found to contain significant quantities of GM_3 associated with the EGF-receptor purified with an antireceptor-sepharose column.[45] In contrast, the EGF receptor purified from the clone A51 cells showed a strong EGF-stimulated cell growth and higher EGF-receptor tyrosine kinase activity, but lacked detectable GM_3.[45] These results further suggest that the association of GM_3 with the EGF receptor leads to an inhibition of its kinase activity. Lyso-GM_3 has also been reported to inhibit tyrosine phosphorylation of the isolated EGF receptor.[45] In contrast to these results, an analog of GM_3 which contains non-N-acetylated neuraminic acid and stimulates cell growth, has been found to stimulate the EGF-receptor tyrosine kinase activity.[34] These results suggest that gangliosides are specific modulators of receptor protein tyrosine kinase activity. The mechanism by which gangliosides modulate this kinase activity is, however, unknown.

If the association of GM_3 with the EGF receptor occurs *in vivo*, and has the same inhibitory effect reported for the addition of exogenous ganglioside, cells should be growth arrested. It has been proposed that GM_3 on the cell surface might be either separated from the receptor or its levels modulated during growth. In this respect, it is interesting to note that plasma membrane-bound sialidase activity has been found to be increased in certain transformed cell lines,[46] and that there is a correlation between cell density with GM_3 metabolism and the activity of two sialidases in the conditioned medium of human fibroblasts.[47]

The exogenous addition of different gangliosides to purified myelin,[48] synaptosomal membranes,[49] or cerebral cortex membranes[50] can both stimulate and inhibit the phosphorylation of certain proteins. The stimulatory pathway presumably may be mediated by the activation of gangliosides of specific protein kinases. The existence of a calcium/ganglioside-dependent protein kinase activity in a neuroblastoma cell line and rat liver has been documented.[41,51] The inhibitory action of gangliosides on protein phosphorylation might be due to activation of protein phosphatases and/or inhibition of protein kinases. Recently, a ganglioside-inhibited protein kinase has been partially purified from guinea pig brain.[52] This protein has an M_r of about 40 kDa and is specifically inhibited by gangliosides, especially polysialogangliosides.[52] Based on the activity of the partially purified enzymes toward various substrates, it has been concluded that this ganglioside-dependent protein kinase is distinct from casein kinases, phosphorylase b kinases, phosvitin kinases, and tyrosine kinases.[52] The nature of the physiological substrates for these kinases is not known and its biological significance has therefore not been determined. However, current results suggest that gangliosides might mediate their biological actions by modulating the activity of novel protein kinases. Fragments of the lipophosphoglycan of *Leishmania donovani*, which were generated by incubation with phospholipase C and mild acid hydrolysis, were found to inhibit rat brain protein kinase C.[78] The 1-*o*-alkylglycerol moiety was the most effective as inhibitor. In addition, several glycolipid antigens of *L. donovani*, which are known to be anchored to the membrane through a phosphatidylinositol anchor (see Chapter 5), were also found to inhibit protein kinase C. The mechanism by which gangliosides, which are present on the cell surface, might affect the activity of intracellular protein kinases is not known. Since transmembrane movement of glycolipids is not likely to occur in the absence of specific transporters, possible mechanisms of action of glycolipids involve their specific transport to the cell interior or their metabolism on the cell surface to yield other molecules (lyso-glycolipids, sphingosine, oligosaccharides, etc.) which might be able to cross the plasma membrane barrier (see below and Chapter 8).

Ganglioside GQ1b has been reported to stimulate ecto-protein kinase activity on the cell surface of a human neuroblastoma cell line, GOTO.[77] Exogenously added gangliosides specifically stimulated the phosphorylation of threonine and serine residues of at least three cell surface-associated proteins with M_rs of 64, 60, and 54 kDa, respectively. GQ1b, at 5 nM, was the most potent among the several gangliosides tested. These results suggest that modulation of the phosphorylation of cell surface proteins might be one of the mechanisms by which gangliosides affect cellular processes such as cell growth and differentiation.

III. BIOLOGICAL FUNCTIONS OF SPHINGOSINE

Sphingosine and other naturally occurring long-chain sphingoid bases inhibit, at low concentrations, certain cellular functions. Preincubation with sphingosine inhibits insulin-stimulated 2-deoxyglucose transport in 3T3-L1 fibroblasts[53] and adipocytes[74] as well as the lipogenic effect of this hormone.[81] Sphingosine also inhibits the differentiation and growth of human promyelocytic leukemic (HL-60) cells induced by phorbol esters[54] and inhibits nerve growth factor-dependent growth in PC-12 cells.[55] Preincubation of human platelets with sphingosine has been reported to block human platelet secretion and second-phase aggregation in response to ADP, thrombin, collagen, arachidonic acid, and platelet-activating factor.[56] Sphingosine had no effect on the initial shape change of platelets or on the first-phase aggregation in response to these agonists. Moreover, platelet aggregation induced by ristocetin was not inhibited by preincubation with sphingosine.[56] Similarly, sphingosine inhibits the generation of superoxide and the secretion of specific, but not azurophilic, granules induced by phorbol esters, the synthesis of platelet-activating factor and leukotriene B$_4$ in neutrophils,[57-59] and the release of arachidonic acid and prostaglandin synthesis in mouse peritoneal macrophages.[82] In human neutrophils, the secretion of lactoferrin by a variety of agonists, including phorbol esters, the chemotactic peptide *N*-formyl-methionyl-leucyl-phenylalanine, and the ionophore A23187, was completely inhibited by sphingosine.[58] Whereas the secretion of lysozyme induced by phorbol esters was completely inhibited, the secretion of this enzyme induced by *N*-formyl-methionyl-leucyl-phenylalanine was only blocked by about 50%.[58] These results suggest that sphingosine interferes with the process of granule secretion in human neutrophils by acting at more than one level. The effects of

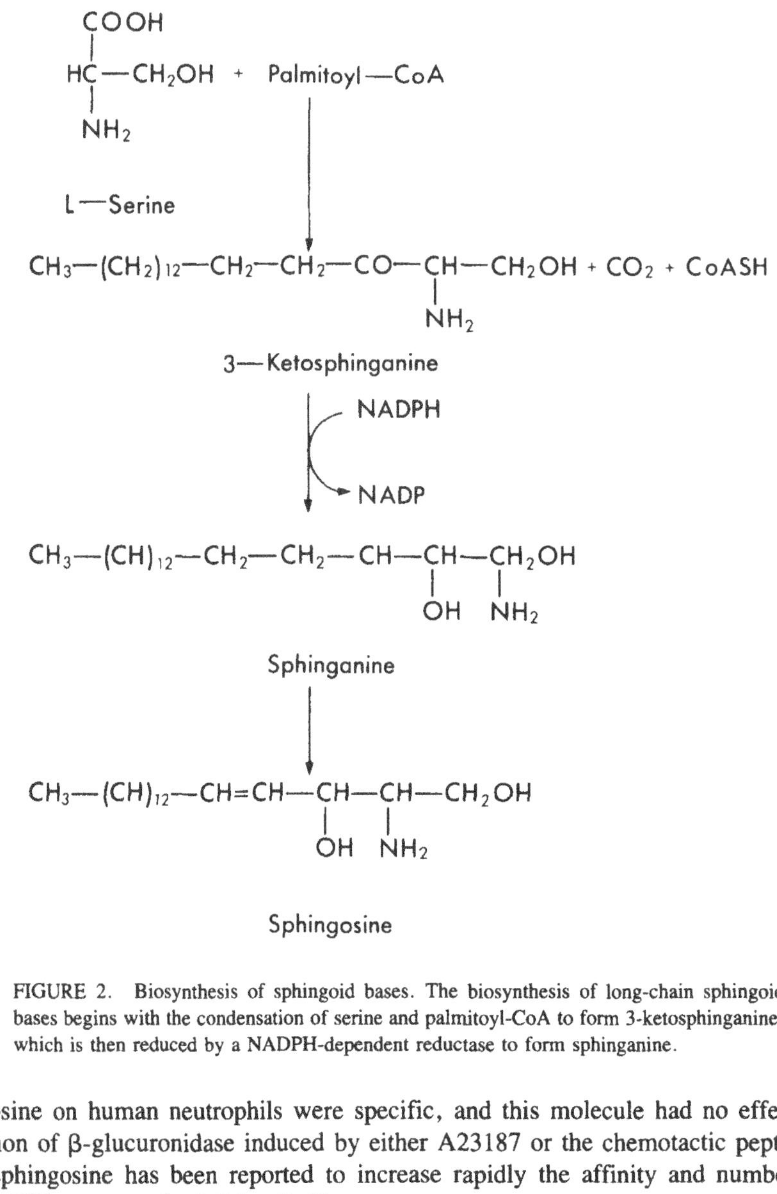

$$\underset{\substack{| \\ NH_2}}{\overset{\substack{COOH \\ |}}{HC}}-CH_2OH \quad + \quad Palmitoyl-CoA$$

L—Serine

$$CH_3-(CH_2)_{12}-CH_2-CH_2-CO-\underset{\substack{| \\ NH_2}}{CH}-CH_2OH + CO_2 + CoASH$$

3—Ketosphinganine

NADPH

NADP

$$CH_3-(CH)_{12}-CH_2-CH_2-\underset{\substack{| \\ OH}}{CH}-\underset{\substack{| \\ NH_2}}{CH}-CH_2OH$$

Sphinganine

$$CH_3-(CH)_{12}-CH=CH-\underset{\substack{| \\ OH}}{CH}-\underset{\substack{| \\ NH_2}}{CH}-CH_2OH$$

Sphingosine

FIGURE 2.   Biosynthesis of sphingoid bases. The biosynthesis of long-chain sphingoid bases begins with the condensation of serine and palmitoyl-CoA to form 3-ketosphinganine, which is then reduced by a NADPH-dependent reductase to form sphinganine.

sphingosine on human neutrophils were specific, and this molecule had no effect on the generation of β-glucuronidase induced by either A23187 or the chemotactic peptide.[58] Finally, sphingosine has been reported to increase rapidly the affinity and number of cell surface EGF receptors in A431 cells.[60]

The enzymatic synthesis of sphingosine was first demonstrated by Brady and Koval, using a cell-free preparation of rat brain and [U-14C]serine.[61] The biosynthesis of long-chain sphingoid bases (see Figure 2) begins with the condensation of serine and palmitoyl-CoA to form 3-ketosphinganine, which is then reduced by a NADPH-dependent reductase to form sphinganine.[62] Sphinganine is then converted to sphingosine by dehydrogenation, through a mechanism which is not well understood.[62] The condensation of serine and palmitoyl-CoA is catalyzed by serine palmitoyltransferase, a pyridoxal 5'-phosphate-dependent enzyme specific for this saturated fatty acid. A variety of results indicate that this reaction is the rate-limiting step in the synthesis of sphingosine. These include the observation that long-chain bases do not accumulate as biosynthetic intermediates and that the activity of the enzyme correlates with the approximate sphingolipid composition of tissues.[63,64] The double bond of sphingosine is probably introduced after the formation *de novo* of N-acylsphinganine.[65,66] Ceramide synthetases can convert sphinganine and other long-chain bases to N-acyl derivatives that are then used for the synthesis of sphingomyelin and glycolipids.[62] The major pathway for the synthesis of sphygomyelin in a number of biological systems is the sphingomyelin synthase reaction. This enzyme transfers the phosphorylcholine headgroup from a molecule of phosphatidylcholine to ceramide, yielding sphingomyelin and diacylglycerol.[62] Many aspects of the enzymology and metabolism of long-chain sphingoid bases, however, remain undetermined.

Low concentrations of long-chain sphingoid bases are thought to exist free in the cell as intermediates of the biosynthesis and degradation of sphingolipids. There are, however, few determinations of the cellular levels of long-chain sphingoid bases. In HL-60 leukemic cells, sphingosine is the major long-chain base detected (about 80% of the total) and is present at about 12.3 pmol/10^6 cells, which represents about 0.2% of the total long-chain bases present in the cells.[54] In human neutrophils, sphingosine levels have been reported to vary from 13 to 101 pmol/10^7 cells for different donors.[67] In these cells, sphingosine also represented about 80% of the total long-chain bases. A continuous increase with time for about 1 h has been observed by incubating human neutrophils at 37°C.[67] This time- and dose-dependent increase in the levels of free sphingosine do not seem to arise from *de novo* pathways, but probably originate from the hydrolysis of other, more complex sphingolipids.[67] A variety of signals which are known to activate human neutrophils, such as phorbol esters, the ionophore A23187, and opsonized zymosan, decreased the time course of sphingosine increase when cells were incubated at 37°C.[67] At low concentrations, the chemotactic peptide N-formyl-methionyl-leucyl-phenylalanine also reduced sphingosine accumulation, but at micromolar concentrations, which are commonly used to activate neutrophils, this molecule increased sphingosine accumulation.[67] Plasma and serum also stimulated sphingosine formation, suggesting that this process occurs in circulating neutrophils.[67] Since sphingosine has been shown to modulate a variety of cellular processes in neutrophils, these results suggest the existence of mechanisms that, by controlling the concentration of free sphingosine, regulate the sensitivity of the cells to various signals. Diacylglycerol, a metabolite commonly formed during cell activation (see Chapter 7), has been shown to stimulate sphingomyelin hydrolysis in GH$_3$ pituitary cells and to cause an increase in ceramide concentration.[68] The concentration of free sphingosine was not determined in these experiments. These results have been interpreted as indicative of an increased hydrolysis of sphingomyelin during GH$_3$ activation,[68] and further suggest the existence of mechanisms that control the levels of sphingosine metabolites during cell stimulation. It is interesting to note that in rat liver, the plasma membrane is the exclusive site of sphingomyelin biosynthesis[69,70] and, hence, this process might be regulated by signals acting on the cell surface. It is not known whether sphingomyelin turnover in other systems is stimulated during cell activation. Changes in the plasma membrane concentration of sphingomyelin, in addition to its possible contribution to changes in the levels of sphingosine and other cellular modulators, can alter the distribution of cholesterol between plasma membranes and intracellular cholesterol pools,[71] which further points to the sphingolipids as important components during cellular signaling.

IV. MECHANISM OF ACTION OF SPHINGOSINE

Sphingosine has been shown to inhibit protein kinase C, using a Triton X-100 mixed micelle assay containing phosphatidylserine and *sn*-1,2-dioctanoylglycerol.[72] The activity of the kinase, which is independent of Ca^{2+}, diacylglycerol, and phospholipids, was not inhibited by sphingosine.[72] Moreover, the effect of sphingosine was subject to surface dilution.[72] These results suggest that sphingosine interacts with the surface-bound protein kinase C, probably by interacting with the regulatory domain of the enzyme without interfering with the active site.[72] The inhibition by sphingosine could be overcome by the addition of phorbol dibutyrate, a cell-permeable activator of protein kinase C, and sphingosine has been shown to inhibit phorbol dibutyrate binding.[72] Moreover, kinetic studies indicated a competitive type of inhibition with respect to diacylglycerol.[72] Studies with a series of long-chain analogs of sphingosine suggest that the major structural features required for inhibition of protein kinase C are the presence of a free amino group and an aliphatic side chain.[83] Sphingosine and related lysophospholipids have also been shown to inhibit the phosphorylation of a 40-kDa protein (substrate of protein kinase C) in human platelets induced by thrombin, phorbol esters, and the cell-permeable protein kinase C activator 1,2-dioctanoyl-glycerol.[17,72] Similarly, sphingosine has also been shown to inhibit phorbol ester-dependent differentiation of the human promyelocytic cell line HL-60,[54] and the inhibition by sphingosine of secondary aggregation was overcome by phorbol dibutyrate and dioctanoylglycerol.[56] In A431 cells, the addition of sphingosine also caused the inhibition of the phosphorylation of two protein substrates of protein kinase C, EGF receptor and transferrin receptor.[60] However, additional experiments indicated that sphingosine also modulates the EGF receptor function by a protein kinase C-independent mechanism.[60] Although sphingosine caused the inhibition of the phosphorylation of the EGF receptor threonine 654, this molecule also stimulated the phosphorylation in theonine of a unique tryptic peptide and produced an increase in the affinity of the receptor under conditions where the stoichiometry of threonine 654 phosphorylation was low.[60] Moreover, the replacement of threonine 654 by an alanine residue did not significantly affect the regulation of EGF binding to its receptor by sphingosine.[60] Finally, sphingomyelinase action, like the sphingoid bases, has also been shown to inhibit certain biological processes mediated via phorbol ester activation of protein kinase C,[84] and sphingomyelin synthase action generates diacylglycerol (see above), which might result in the stimulation of protein kinase C.[85]

Like sphingosine, many calmodulin antagonists contain a hydrophobic moiety and a positively charged amino group. The inhibitory effect of sphingosine on calmodulin has been studied using the multifunctional Ca^{2+}/calmodulin-dependent protein kinase. Sphingosine has been shown to cause a dose-dependent inhibition of the phosphorylation of microtubule-associated protein 2 by the Ca^{2+}/calmodulin-dependent kinase.[73] The results obtained suggest that micromolar concentrations of sphingosine interfere with the ability of calmodulin to activate the enzyme, and that sphingosine is competitive with calmodulin.[73] Sphingosine seems to be a more general inhibitor of calmodulin actions, since it inhibited the activation by calmodulin of smooth-muscle myosin light-chain kinase and of the Ca^{2+}/calmodulin-dependent phosphodiesterase.[73] Finally, in a pituitary cell line, GH_3 sphingosine has been shown to inhibit the phosphorylation of the microtubule-associated protein 2 by the multifunctional Ca^{2+}/calmodulin-dependent protein kinase and the phosphorylation of elongation factor 2 by the Ca^{2+}/calmodulin-dependent protein kinase III.[73]

In conclusion, the exogenous addition of sphingolipids modulates a variety of cellular functions. Different results indicate that, in the micromolar range, sphingolipids are potent inhibitors of a variety of protein kinases *in vitro*, especially of protein kinase C, and might also regulate these enzymes *in vivo*. As mentioned in Chapter 7, the hydrolysis of phosphoinositides during cell signaling produces two types of second messengers: diacylglycerol and inositol-P_3, which, in turn, releases Ca^{2+} from intracellular stores. Sphingolipids might be negative effectors of the activation of protein kinase C by diacylglycerol and Ca^{2+}, and of the Ca^{2+}/calmodulin-dependent protein kinase by Ca^{2+} and calmodulin. To determine the physiological significance of sphingolipids in cellular signaling, it will be necessary to determine the basal levels of these molecules and their variations in response to different signals. As mentioned above, there is only limited information about the effect of several agonists on sphingolipid turnover.

REFERENCES

1. **Pick, L. and Bielschowsky, M.**, Über lipoidzellige splenomegalie (typus Niemann-Pick) und amaurotische idiotie, *Klin. Wochenschr.*, 6, 1631, 1927.
2. **Shapiro, D. and Flowers, H. M.**, Studies on sphingolipids. VII. Synthesis and configuration of natural sphingomyelins, *J. Am. Chem. Soc.*, 84, 1047, 1962.
3. **Sweeley, E. L. and Moscatelli, E. A.**, Qualitative microanalysis and estimation of sphingolipid bases, *J. Lipid Res.*, 1, 40, 1959.
4. **Hirvisalo, E. L. and Renkonen, O.**, Composition of human serum sphingomyelins, *J. Lipid Res.*, 11, 54, 1970.
5. **Samuelsson, B. and Samuelsson, K.**, Separation and identification of ceramides derived from human plasma sphingomyelins, *J. Lipid Res.*, 10, 47, 1969.
6. **Karlsson, K. A. and Steen, G. O.**, Studies on sphingosines. XIII. The existence of phytosphingosine in bovine kidney sphingomyelins, *Biochim. Biophys. Acta*, 152, 789, 1968.
7. **Svennerholm, E., Stallberg-Stenhagen, S., and Svennerholm, L.**, Fatty acid composition of sphingomyelins in blood, spleen, placenta, liver, lung and kidney, *Biochim. Biophys. Acta*, 125, 60, 1966.
8. **O'Brien, J. S. and Rouser, G.**, The fatty acid composition of brain sphingolipids: sphingomyelin, ceramide, cerebroside, and cerebroside sulfate, *J. Lipid Res.*, 5, 339, 1964.
9. **Merrill, A. H., Wang, E., and Wertz, P. W.**, Differences in the long chain (sphingoid) base composition of sphingomyelin from rats bearing Morris hepatoma 7777, *Lipids*, 21, 529, 1986.
10. **Hakomori, S.**, Glycosphingolipids in cellular interaction, differentiation and oncogenesis, in *Handbook of Lipid Research: Sphingolipid Biochemistry*, Vol. 3, Hannahan, D. J., Ed., Academic Press, New York, 1983, 327.
11. **Wiegandt, H.**, Gangliosides, in *Glycosphingolipids*, Wiegandt, H., Ed., Elsevier, Amsterdam, 1985, 199.
12. **Steck, T. L. and Dawson, G.**, Topographical distribution of complex carbohydrates of the erythrocyte membrane, *J. Biol. Chem.*, 249, 2135, 1974.
13. **Hedo, J. A. and Kahn, R. C.**, Radioactive labeling and turnover studies of the insulin receptor subunits, *Methods Enzymol.*, 109, 593, 1985.
14. **Hakomori, S.**, Glycosphingolipids in cellular interaction, differentiation, and oncogenesis, *Annu. Rev. Biochem.*, 50, 733, 1981.
15. **Symington, F., Murray, W. A., Bearman, S. I., and Hakomori, S.**, Intracellular localization of lactosylceramide, the major human neutrophil glycosphingoid, *J. Biol. Chem.*, 262, 11356, 1987.
16. **Sharom, F. J. and Grant, C. W. M.**, A model for ganglioside behavior in cell membranes, *Biochim. Biophys. Acta*, 507, 208, 1978.
17. **Hannun, Y. A. and Bell, R. M.**, Functions of sphingolipids and sphingolipid breakdown products in cellular regulation, *Science*, 243, 500, 1989.
18. **Gahmberg, C. G. and Hakomori, S.**, Organization of glycolipids and glycoproteins in surface membranes: dependency on cell cycle and on transformation, *Biochem. Biophys. Res. Commun.*, 59, 283, 1974.
19. **Gahmberg, C. G. and Hakomori, S.**, Surface carbohydrates of hamster fibroblasts, *J. Biol. Chem.*, 250, 2438, 1975.
20. **Fishman, P. H. and Brady, R. O.**, Biosynthesis and functions of gangliosides, *Science*, 194, 906, 1976.
21. **Matyas, G. R., Aaronson, S. A., Brady, R. O., and Fishman, P. H.**, Alteration of glycolipids in *ras*-transfected NIH 3T3 cells, *Proc. Natl. Acad. Sci. U.S.A.*, 84, 6065, 1987.
22. **Bremer, E. G., Hakomori, S., Bowen-Pope, D. F., Raines, E., and Ross, R.**, Ganglioside-mediated modulation of cell growth, growth factor binding, and receptor phosphorylation, *J. Biol. Chem.*, 259, 6818, 1984.
23. **Hanai, N., Dohi, T., Nores, G. A., and Hakomori, S.**, A novel ganglioside, de-N-acetyl-GM₃ (neuraminyllactosylceramide, II3NeuNH₂LacCer), acting as a strong promoter for epidermal growth factor receptor kinase and as a stimulator for cell growth, *J. Biol. Chem.*, 263, 6296, 1988.
24. **Fishman, P. H., Simmons, J. L., Brady, R. O., and Freese, E.**, Induction of glycolipid biosynthesis by sodium butyrate, *Biochem. Biophys. Res. Commun.*, 59, 292, 1974.
25. **Patt, L., Itaya, K., and Hakomori, S.**, Retinol induces density-dependent growth inhibition and changes in glycolipids, *Nature*, 273, 379, 1978.
26. **Burczak, J. D., Moskal, J. R., Trosko, J. E., Fairley, J. L., and Sweeley, C. C.**, Phorbol ester-associated changes in ganglioside metabolism, *Exp. Cell Res.*, 147, 281, 1983.
27. **Laine, R. A. and Hakomori, S.**, Incorporation of exogenous glycosphingolipids in plasma membranes of cultured hamster cells and concurrent change of growth behavior, *Biochem. Biophys. Res. Commun.*, 54, 1039, 1973.
28. **Keenan, T. W., Schmid, E., Franke, W. W., and Wiegandt, H.**, Exogenous glycosphingolipids suppress growth rate of transformed and untransformed 3T3 mouse cells, *Exp. Cell Res.*, 92, 259, 1975.

29. **Bremer, E. G. and Hakomori, S.**, GM₃ ganglioside induces hamster fibroblast growth inhibition on chemically-defined medium: ganglioside may regulate growth factor receptor function, *Biochem. Biophys. Res. Commun.*, 106, 711, 1982.
30. **Spiegel, S.**, Fluorescent derivatives of ganglioside GM1 function as receptor for cholera toxin, *Biochemistry*, 24, 5947, 1985.
31. **Spiegel, S. and Fishman, P. H.**, Gangliosides as bimodal regulators of cell growth, *Proc. Natl. Acad. Sci. U.S.A.*, 84, 141, 1987.
32. **Spiegel, S. and Panagiotopoulos, C.**, Mitogenesis of 3T3 fibroblasts induced by endogenous ganglioside is not mediated by cAMP, protein kinase C, or phosphoinositides turnover, *Exp. Cell Res.*, 177, 414, 1988.
33. **Facci, L., Skaper, S. D., Favaron, M., and Leon, A.**, A role of gangliosides in astroglial cell differentiation in vitro, *J. Cell Biol.*, 106, 821, 1988.
34. **Igarashi, Y., Nojiri, H., Hanai, N., and Hakomori, S.**, Gangliosides that modulate membrane protein function, *Methods Enzymol.*, 179, 521, 1989.
35. **Ledeen, R. W.**, Gangliosides of the neuron, *Trends Neurosci.*, 8, 169, 1985.
36. **Gorio, A., Carmignoto, G., Facci, L., and Finesso, M.**, Motor sprouting induced by ganglioside treatment. Possible implication for gangliosides on neuronal growth, *Brain Res.*, 197, 236, 1980.
37. **Agnati, L. F., Fuxe, K., Calza, L., Benfenati, F., Cavicchioli, L., Toffano, G., and Goldstein, M.**, Gangliosides increase the survival of lesioned dopamine neurons and favour the recovery of dopaminergic synaptic function in striatum of rats by collateral sprouting, *Acta Physiol. Scand.*, 119, 347, 1983.
38. **Agnati, L. E., Benfenati, F., Battistini, N., Cavicchioli, L., Fuxe, K., and Toffano, G.**, Selective modulation of ³H-spiperone labelled 5-HT receptors by subchronic treatment with the ganglioside GM₁ in the rat, *Acta Physiol. Scand.*, 117, 311, 1983.
39. **Agnati, L. F., Fuxe, K., Benfenati, F., Battistini, N., Zini, I., and Toffano, G.**, Chronic ganglioside treatment counteracts the biochemical signs of dopamine receptor supersensitivity induced by chronic haloperidol treatment, *Neurosci. Lett.*, 40, 293, 1983.
40. **Cuello, A. C., Garofalo, L., Kenigsberg, R. L., and Maysinger, D.**, Gangliosides potentiate *in vivo* and *in vitro* effects of nerve growth factor on central cholinergic neurons, *Proc. Natl. Acad. Sci. U.S.A.*, 86, 2056, 1989.
41. **Tsuji, S., Nakajima, J., Sasaki, T., and Nagai, J.**, Bioactive gangliosides. IV. Ganglioside GO₁ᵦ/Ca²⁺-dependent protein kinase activity exists in the plasma membrane fraction of neuroblastoma cell line, GOTO, *J. Biochem.*, 97, 969, 1985.
42. **Dixon, S. J., Stewart, D., Grinstein, S., and Spiegel, S.**, Transmembrane signaling by the B subunit of cholera toxin: increased cytoplasmic free calcium in rat lymphocytes, *J. Cell Biol.*, 105, 1153, 1987.
43. **Spiegel, S.**, Possible involvement of a GTP-binding protein in a late event during endogenous ganglioside-modulated cellular proliferation, *J. Biol. Chem.*, 264, 6766, 1989.
44. **Bremer, E. G., Schlessinger, J., and Hakomori, S.**, Ganglioside-mediated modulation of cell growth. Specific effects of GM3 on tyrosine phosphorylation of the epidermal growth factor receptor, *J. Biol. Chem.*, 261, 2434, 1986.
45. **Hanai, N., Nores, G. A., MacLeod, C., Torres-Mendez, C. R., and Hakomori, S.**, Ganglioside-mediated modulation of cell growth. Specific effects of GM₃ and lyso-GM₃ in tyrosine phosphorylation of the epidermal growth factor receptor, *J. Biol. Chem.*, 263, 10915, 1988.
46. **Schengrund, C. L., Lausch, R. N., and Rosenberg, A.**, Sialidase activity in transformed cells, *J. Biol. Chem.*, 248, 4424, 1973.
47. **Usuki, S., Lyu, S. C., and Sweeley, C. C.**, Sialidase activities of cultured human fibroblasts and the metabolism of GM₃ ganglioside, *J. Biol. Chem.*, 263, 6847, 1988.
48. **Chan, K. F. J.**, Ganglioside-modulated protein phosphorylation in myelin, *J. Biol. Chem.*, 262, 2415, 1987.
49. **Chan, K. F. J.**, Ganglioside-modulated protein phosphorylation. Partial purification and characterization of a ganglioside-stimulated protein kinase in brain, *J. Biol. Chem.*, 262, 5248, 1987.
50. **Cimino, M., Benfenati, F., Farabegoli, C., Cattabeni, F., Fuxe, K., Agnati, L. F., and Toffano, G.**, Differential effect of ganglioside GM₁ on rat brain phosphoproteins: potentiation and inhibition of protein phosphorylation regulated by calcium/calmodulin and calcium/phospholipid-dependent protein kinases, *Acta Physiol. Scand.*, 130, 317, 1987.
51. **Goldenring, J. R., Otis, L. C., Yu, R. K., and De Lorenzo, R. J.**, Calcium/ganglioside-dependent protein kinase activity in rat brain membrane, *J. Neurochem.*, 44, 1229, 1985.
52. **Chan, K. F. J.**, Ganglioside-modulated protein phosphorylation. Partial purification and characterization of a ganglioside-inhibited protein kinase in brain, *J. Biol. Chem.*, 263, 568, 1988.
53. **Nelson, D. and Murray, D.**, Sphingolipids inhibit insulin and phorbol ester stimulated uptake of 2-deoxyglycose, *Biochem. Biophys. Res. Commun.*, 138, 463, 1986.
54. **Merrill, A., Sereni, A. M., Stevens, V. L., Hannun, Y. A., Bell, R. M., and Kinkade, J. M.**, Inhibition of phorbol ester-dependent differentiation of human promyelocytic leukaemic (HL-60) cells by sphinganine and other long-chain bases, *J. Biol. Chem.*, 261, 12610, 1986.

55. **Hall, F. L., Fernyhough, P., Ishii, D. N., and Vulliet, P. R.,** Suppression of nerve growth factor-derived neurite outgrowth in PC12 cells by sphingosine, and inhibitors of protein kinase C, *J. Biol. Chem.*, 263, 4460, 1988.

56. **Hannun, Y., Greenberg, C. S., and Bell, R. M.,** Sphingosine inhibition of agonist-dependent secretion and activation of human platelets implies that protein kinase C is a necessary and common event of the signal transduction pathways, *J. Biol. Chem.*, 262, 13620, 1987.

57. **Wilson, E., Olcott, M. C., Bell, R. M., Merrill, A. H., and Lambeth, J. D.,** Inhibition of the oxidative burst in human neutrophils by sphingoid long-chain bases. Role of protein kinase C inactivation of the burst, *J. Biol. Chem.*, 261, 12616, 1986.

58. **Wilson, E., Rice, W. G., Kinkade, J. M., Merrill, A. H., Arnold, R. R., and Lambeth, J. D.,** Protein kinase C inhibition by sphingoid long-chain bases: effects on secretion in human neutrophils, *Arch. Biochem. Biophys.*, 259, 204, 1987.

59. **McIntyre, T. M., Reihold, S. L., Prescott, G. A., and Zimmerman, G. A.,** Protein kinase C activity appears to be required for the synthesis of platelet activating factor and leukotriene B_4 by human neutrophils, *J. Biol. Chem.*, 262, 15370, 1987.

60. **Faucher, M., Girones, N., Hannun, Y. A., Bell, R., and Davis, R. J.,** Regulation of the epidermal growth factor receptor phosphorylation state by sphingosine in A431 human epidermoid carcinoma cells, *J. Biol. Chem.*, 263, 5319, 1988.

61. **Brady, R. O. and Koval, G. J.,** The enzymatic synthesis of sphingosine, *J. Biol. Chem.*, 233, 26, 1958.

62. **Kishimoto, Y.,** Sphingolipid formation, in The Enzymes, Vol. 16, Boyer, P. B., Ed., Academic Press, New York, 1983, 357.

63. **Merrill, A. H. and Wang, E.,** Biosynthesis of long-chain (sphingoid) bases from serine by LM cells. Evidence for introduction of the C-trans-double bond after de novo biosynthesis of N-acyl sphinganine(s), *J. Biol. Chem.*, 261, 3764, 1986.

64. **Merrill, A. H., Wang, E., and Mullins, R. E.,** Kinetics of long-chain (sphingoid) base biosynthesis in intact LM cells: effects of varying the extracellular concentrations of serine and fatty acid precursors of this pathway, *Biochemistry*, 27, 340, 1988.

65. **Ong, D. E. and Brady, R. N.,** In vivo studies on the introduction of the 4-trans-double bond of the sphinganine moiety of rat brain ceramides, *J. Biol. Chem.*, 248, 3884, 1973.

66. **Merrill, A. H. and Wang, E.,** Biosynthesis of long-chain (sphingoid) bases from serine by LM cells. Evidence for introduction of the 4-*trans*-double bond after *de novo* biosynthesis, *J. Biol. Chem.*, 261, 3764, 1986.

67. **Wilson, E., Wang, E., Mullins, R. E., Uhlinger, D. J., Liotta, D. C., Lambeth, J. D., and Merrill, A. H.,** Modulation of the free sphingosine levels in human neutrophils by phorbol esters and other factors, *J. Biol. Chem.*, 263, 9304, 1988.

68. **Kolesnick, R. N.,** 1,2-Diacylglycerols but not phorbol esters stimulate sphingomyelin hydrolysis in GH_3 pituitary cells, *J. Biol. Chem.*, 262, 16759, 1987.

69. **Voelker, D. R., Kennedy, E. P.,** Cellular and enzymic synthesis of sphingomyelin, *Biochemistry*, 21, 2753, 1982.

70. **van den Hill, A., Heusden, P. H., and Wirtz, K. W. A.,** The synthesis of sphingomyelin in the Morris hepatomas 7777 and 5123D is restricted to the plasma membrane, *Biochim. Biophys. Acta*, 833, 354, 1985.

71. **Slotte, J. P. and Bierman, E. L.,** Depletion of plasma-membrane sphingomyelin rapidly alters the distribution of cholesterol between plasma membranes and intracellular cholesterol pools in cultured fibroblasts, *Biochem. J.*, 250, 653, 1988.

72. **Hannun, Y. A., Loomis, C. R., Merrill, A. H., and Bell, R. M.,** Sphingosine inhibition of protein kinase C activity and of phorbol dibutyrate *in vitro* and in human platelets, *J. Biol. Chem.*, 261, 12604, 1986.

73. **Jefferson, A. B. and Schulman, H.,** Sphingosine inhibits calmodulin dependent enzymes, *J. Biol. Chem.*, 263, 15241, 1988.

74. **Robertson, D. G., DiGirolamo, M., Merrill, A. H., and Lambeth, J. D.,** Insulin-stimulated hexose transport and glucose oxidation in rat adipocytes is inhibited by sphingosine at a step after insulin binding, *J. Biol. Chem.*, 264, 6778, 1989.

75. **DeGrandis, S., Law, H., Brunton, J., Gyles, C., and Lingwood, C. A.,** Globotetraosylceramide is recognized by the pig edema disease toxin, *J. Biol. Chem.*, 264, 12520, 1989.

76. **Bansal, R. and Pfeiffer, S. E.,** Reversible inhibition of oligodendrocyte progenitor differentiation by a monoclonal antibody against surface galactolipids, *Proc. Natl. Acad. Sci. U.S.A.*, 86, 6181, 1989.

77. **Tsuji, S., Yamashita, T., and Nagai, Y.,** A novel, carbohydrate signal-mediated cell surface protein phosphorylation: ganglioside GQ1b stimulates ecto-protein kinase activity on the cell surface of a human neuroblastoma cell line, GOTO, *J. Biochem.*, 104, 498, 1988.

78. **McNealy, T. B., Rosen, G., Londner, M. V., and Turco, S. J.,** Inhibitory effects on protein kinase C activity by lipophosphoglycan fragments and glycophosphatidylinositol antigens of the protozoan parasite *Leishmania*, *Biochem. J.*, 259, 601, 1989.

79. **Spiegel, S.,** Inhibition of protein kinase C-dependent cellular proliferation by interaction of endogenous ganglioside GM₁ with the β-subunit of cholera toxin, *J. Biol. Chem.,* 264, 16512, 1989.
80. **Skaper, S. D., Facci, L., Favaron, M., and Leon, A.,** Inhibition of DNA synthesis in C6 glioma cells following cellular incorporation of GM₁ ganglioside and choleragenoid exposure, *J. Neurochem.,* 51, 688, 1988.
81. **Smal, J. and De Meyts, P.,** Sphingosine, an inhibitor of protein kinase C, suppresses the insulin-like effects of growth hormone in rat adipocytes, *Proc. Natl. Acad. Sci. U.S.A.,* 86, 4705, 1989.
82. **Pfannkuche, H. J., Kaever, V., Gemsa, D., and Resch, K.,** Regulation of prostaglandin synthesis by protein kinase C in mouse peritoneal macrophages, *Biochem. J.,* 260, 471, 1989.
83. **Merrill, A. H., Nimkar, S., Menaldino, D., Hannun, Y. A., Loomis, C., Bell, R. M., Tyagi, S. R., Lambeth, J. D., Stevens, V. L., and Hunter, R.,** Structural requirements for long-chain (sphingoid) base inhibition of protein kinase C in vitro and for the cellular effects of these compounds, *Biochemistry,* 28, 3138, 1989.
84. **Kolesnick, R. N.,** Sphingomyelinase action inhibits phorbol ester-induced differentiation of human pro-myelocytic leukemic (HL-60) cells, *J. Biol. Chem.,* 264, 7617, 1989.
85. **Hampton, R. Y. and Morand, O. H.,** Sphingomyelin synthase and PKC activation, *Science,* 246, 1050, 1989.

Chapter 10

BIOLOGICAL ACTIONS OF THE ETHER-LINKED GLYCEROPHOSPHOLIPID PLATELET ACTIVATING FACTOR

José M. Mato

TABLE OF CONTENTS

I. INTRODUCTION

PAF (1-*O*-alkyl-2-acetyl-*sn*-glycero-3-phosphocholine) is a biologically active ether-linked glycerophospholipid that activates platelets at 10^{-10} *M* and seems to serve as a critical mediator in diverse biological processes such as inflammation, anaphylaxis, and neuronal development.[1-3] PAF is also a potent hypotensive agent and it has been shown to induce smooth muscle contraction and bronchoconstriction. Because of its biological activities both *in vitro* and when administered *in vivo*, PAF has been proposed to be implicated in many diseases (e.g., asthma, acute allergic reactions, septic shock, and gastrointestinal ulceration); however, it is not yet clear in which natural pathogenic or physiologic processes this molecule takes part.

In 1970, Henson proposed that leukocytes of sensitized immunized rabbits released a "fluid phase mediator" that induced histamine release from rabbit platelets.[4] This observation was confirmed by other investigators and the term platelet-activating factor (PAF) was coined.[5,6] In 1979, two groups reported the semisynthetic pathway for the synthesis of a 1-*O*-alkyl-2-acetyl-*sn*-glycero-3-phosphocholine (abbreviated PAF, PAF-acether, or AGEPC) which mimicked the effect of PAF on platelets.[7,8] The final structure of this molecule, which coincided with the structure of an hypotensive factor,[9] was published in 1980 (Figure 1).[10] There is evidence favoring the existence of multiple molecular species of PAF in neutrophils which vary in the nature of the alkyl group.[11-14] Alkyl chain homologs of PAF containing C15:0, C16:0, C17:0, C18:0, C18:1, and C22:0 have been isolated from neutrophils,[11-14] although C16:0 and C18:0 seem to be the most abundant species.[15,16] In addition, 1-*O*-acyl analogs of PAF have also been reported.[17,18] Since the various PAF derivatives stimulate platelets to different degrees,[14] these results indicate that neutrophils release a variety of PAF derivatives with various biological potencies and perhaps functions.

II. BIOSYNTHESIS OF PLATELET-ACTIVATING FACTOR

Alkyl and alk-1-enyl glycerolipids were identified and the precise chemical structure of these compounds reported between 1915 and 1924.[19] However, it took about 50 years to establish the biosynthetic pathways involved in the synthesis of these ether-linked glycerophospholipids. Long-chain fatty alcohols are the source of the alkyl group, and this was found to occur through a microsomal reaction in which the long-chain fatty alcohol replaces the acyl group of a molecule of acyl-dihydroxyacetone phosphate (Figure 2).[20-23] The ketone group of alkyl-dihydroxyacetone phosphate is then reduced by an NADPH-linked oxidoreductase to form 1-*O*-alkyl-2-lyso-*sn*-glycero-3-phosphate, the alkyl analog of lyso phosphatidic acid,[21-23] which can then be utilized as a substrate by an acyltransferase to form the alkyl analog of phosphatidic acid which, in turn, can then be utilized by a phosphohydrolase to form 1-*O*-alkyl-2-acyl-*sn*-glycerol.[21,24] Hence, the pathway for the synthesis of 1-*O*-alkyl-2-acyl-*sn*-glycero-3-phosphoethanolamine and 1-*O*-alkyl-2-acyl-*sn*-glycero-3-phosphocholine is analogous to the one established for the diacyl phospholipids by Kennedy and Weiss (see Chapter 6).[25] The utilization of 1-*O*-alkyl-2-acyl-*sn*-glycerols by ethanolamine or choline phosphotransferases, which require cytidine diphosphate derivatives of ethanolamine or choline to form 1-*O*-alkyl-2-acyl-*sn*-glycero-3-phosphoethanolamine or 1-*O*-alkyl-2-acyl-*sn*-glycero-3-phosphocholine, has been established for a variety of tissues, using microsomal preparations.[26-28]

There are two different enzymatic systems for the biosynthesis of PAF. The first system uses an acetyl-CoA:1-*O*-alkyl-2-lyso-glycero-3-phosphocholine acetyltransferase[29] and the second system, a CDP-choline:1-alkyl-2-acetyl-glycerol cholinephosphotransferase (Figure 3).[30,59] The biosynthesis of PAF by both pathways starts with the synthesis of 1-*O*-alkyl-2-lyso-*sn*-glycero-3-phosphate from alkyldihydroxyacetone phosphate by an NADPH-depen-

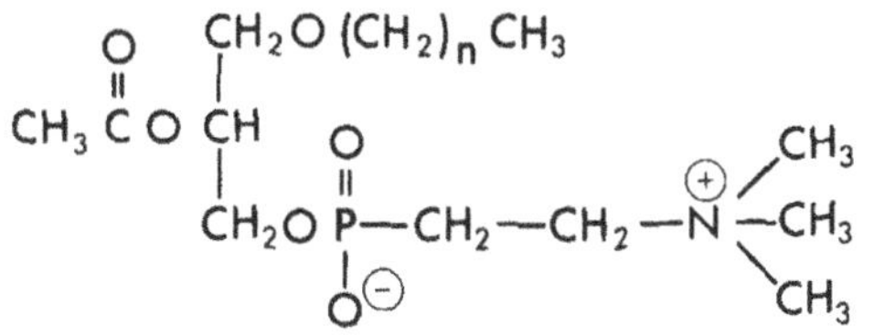

FIGURE 1.   Structure of PAF; 1-*O*-alkyl-2-acetyl-*sn*-glycero-3-phosphocholine. C16:0 and C18:0 seem to be the most abundant species for the alkyl group.

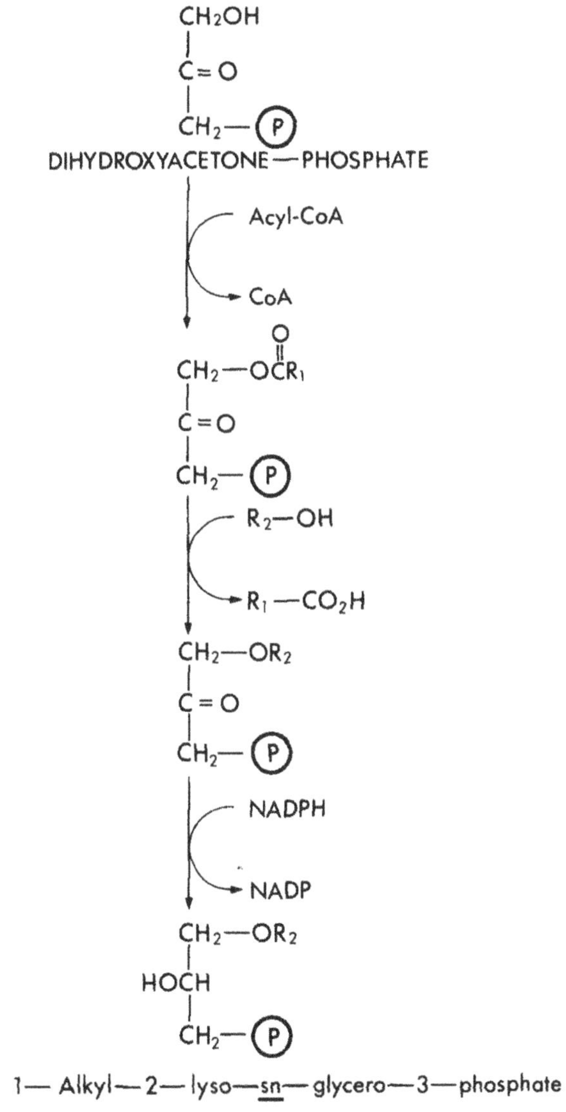

FIGURE 2.   Biosynthesis of 1-*O*-alkyl-2-lyso-*sn*-glycero-3-phosphate from dihydroxyacetone phosphate. In this pathway, a long-chain fatty alcohol replaces the acyl group of a molecule of acyl-dihydroxyacetone phosphate. The ketone group of alkyl-dihydroxyacetone phosphate is then reduced by an NADPH-linked oxidoreductase to form 1-*O*-alkyl-2-lyso-*sn*-glycero-3-phosphate, the alkyl analog of lyso phosphatidic acid.

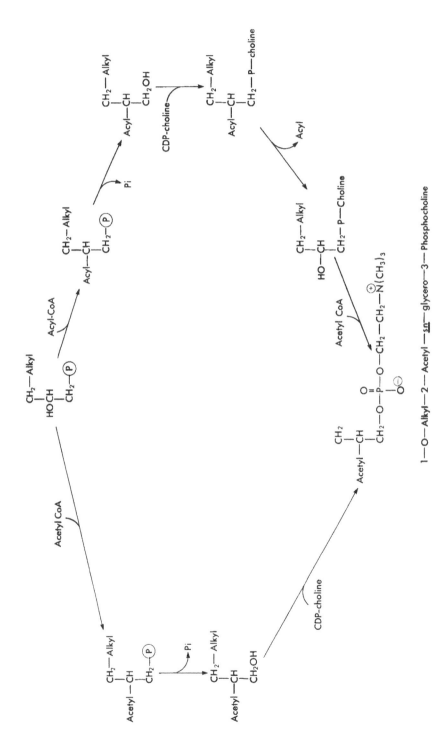

FIGURE 3. Schematic representation of the two pathways for the biosynthesis of platelet-activating factor.

dent oxidoreductase (see above, Figure 2). The lipid generated in this reaction can serve as either a substrate for an acyl-CoA acyltransferase, to form 1-O-alkyl-2-acyl-sn-glycero-3-phosphate, or for an acetyl-CoA acetyltransferase, to form 1-O-alkyl-2-acetyl-sn-glycero-3-phosphate (Figure 3). 1-O-alkyl-2-acetyl-sn-glycero-3-phosphate is then dephosphorylated by a phosphohydrolase[60] and the alkylacylglycerol formed is utilized by a microsomal DTT-insensitive cholinephosphotransferase that catalyzes the final step in the biosynthesis of PAF by this pathway (Figure 3).[1,30,59] In the case of the pathway using 1-O-alkyl-2-acyl-sn-glycero-3-phosphate, this molecule is converted into 1-O-alkyl-2-acyl-sn-glycerol, by the action of a phosphohydrolase, and then to 1-O-alkyl-2-acyl-sn-glycero-3-phosphocholine, by a choline phosphotransferase, as discussed above (Figure 2). This molecule is then converted to 1-O-alkyl-2-lyso-sn-glycero-3-phosphocholine (lyso PAF) by the action of a phospholipase A_2, and, finally, to PAF by the action of a 1-alkyl-2-lyso-sn-glycero-3-phosphocholine acetyltransferase (Figure 3).[1,29]

The purpose for the existence of these two different pathways for the biosynthesis of PAF is not clear. However, this is not an unique situation in lipid metabolism. Similarly, phosphatidylcholine is also synthesized by two different pathways, the CDP-choline pathway and the transmethylation pathway, and phosphatidylethanolamine can be synthesized by the CDP-ethanolamine pathway and from phosphatidylserine (see Chapter 6). The specific activity of these two pathways in producing PAF is different. As a rule, under resting conditions, the specific activity of the DTT-insensitive cholinephosphotransferase is much higher than the specific activity of the 1-O-alkyl-2-lyso-glycero-3-phosphocholine acetyltransferase.[35-37] The pathway involving the acetylation of lyso PAF has been shown to be activated by all agents known to stimulate the synthesis of PAF. Thus, 1-O-alkyl-2-lyso-glycero-3-phosphocholine acetyltransferase is stimulated severalfold in human neutrophils, rat alveolar and peritoneal macrophages, and eosinophils treated with zymosan[38-40] or the calcium ionophore A23187.[40,41] Stimulation of the acetyltransferase reaction preceded the formation and release of PAF by human neutrophils stimulated with zymosan.[38] In peritoneal macrophages from mice injected with thioglycollate, the reduced release of PAF obtained after zymosan addition correlates with a reduction in the activity of the acetyltransferase reaction.[42] The DTT-insensitive cholinephosphotransferase does not seem to be stimulated during cell activation. Thus, this enzyme has been shown to be unaffected after zymosan or thrombin addition to human neutrophils,[40,43] although in Ehrlich ascites cells, DTT-insensitive cholinephosphotransferase is markedly stimulated by the addition of 2 mM oleic acid to the culture media,[61] a situation similar to that observed in the liver for the synthesis of diacyl phosphatidylcholine by the CDP-choline pathway (see Chapter 6). Moreover, alkyl-dihydroxyacetone phosphate synthase activity is also unaffected after stimulation with the ionophore A23187 of human neutrophils.[40] Based on these observations, it has been proposed that the DTT-insensitive cholinephosphotransferase pathway (also called the de novo pathway) is necessary to maintain physiological levels of PAF under resting conditions, whereas the 1-O-alkyl-2-lyso-glycero-3-phosphocholine acetyltransferase pathway is responsible for the biosynthesis of PAF during a stimulatory response.[44] Finally, since alkylacylglycerophosphocholines are rich in polyunsaturated fatty acids (see Chapter 2), when these molecules are converted by the action of a phospholipase A_2 to lyso-PAF, then to be acetylated to form PAF, the fatty acid generated is predominantly arachidonic acid and docosahexaenoic acid.[51-55] Therefore, by the glycero-3-phosphocholine acetyltransferase pathway, two different types of lipid mediators are generated, PAF and arachidonic acid, the later being a precursor for the biosynthesis of prostaglandins and leukotrienes.

The mechanism by which the various signals stimulate the acetylation of lyso-PAF is not well understood, although there is evidence in favor of a phosphorylation-dephosphorylation process. Acetyltransferase activity of spleen microsomes is stimulated by incubation with Mg^{2+}/ATP, and incubation with alkaline phosphatase reduces its activity.[45] The addition of phosphatidylserine or diolein, but not exogenous calmodulin, produced a further stimulation of the acetylation reaction.[45] These results have been interpreted as indicative of a role of protein kinase C in the modulation of 1-O-alkyl-2-lyso-sn-glycero-3-phosphocholine acetyltransferase.[45] However, the addition of partially purified protein kinase C to homogenates from ionophore A23187-stimulated human polymorphonuclear leukocytes reduced acetyltransferase activity, suggesting that in this system protein kinase C is not involved in the activation, but perhaps in the inactivation, of the acetylation reaction.[46] More recently, the addition of phorbol myristate to human neutrophils, but not of its inactive analog 4-alpha-phorbol-12,13-didecanoate, has been shown to stimulate lyso-PAF production, followed by acetyltransferase activation and generation of PAF.[88] The protein kinase C inhibitor, sphingosine (see Chapter 9), inhibited the generation of lyso-PAF and PAF and the stimulation of acetyltransferase induced by the phorbol ester or opsonized zymosan.[88] Similarly, pretreatment of endothelial cells with phorbol esters that activate protein kinase C, or with dioctanoylglycerol followed by stimulation with agonists that induce PAF generation, resulted in a twofold increase in the production of PAF.[89] These results strongly suggest the involvement of protein kinase C in PAF production. Incubation of spleen microsomes with Mg^{2+}/ATP and the catalytic subunit of cAMP-dependent protein kinase increased acetyltransferase activity up to threefold.[47] The nonhydrolyzable ATP derivative AMP-PNP could not replace ATP for the activation of acetyltransferase, and the heat-stable inhibitor of the cAMP-dependent protein kinase prevented activation of the enzyme.[47] Similar results have been observed with homogenates from human polymorphonuclear cells and mouse mast cells.[46,56] 1-O-alkyl-2-lyso-sn-glycero-3-phosphocholine acetyltransferase has been partially purified from rat spleen microsomes.[48] Incubation of the partially purified enzyme with 1-O-[^{3}H]octadecyl-2-lyso-sn-glycero-3-phosphocholine followed by electrophoresis resulted in the incorporation of radioactivity into a protein with an M_r of about 30 kDa.[48] This partially purified enzyme was activated by incubation with the catalytic subunit of cAMP-dependent protein kinase, and activation coincided with the incorporation of [^{32}P] into a 30-kDa protein band, as analyzed by SDS-polyacrylamide gel electrophoresis.[49] Although these results are suggestive of a phosphorylation/dephosphorylation mechanism of control of the activity of the acetylation reaction in the synthesis of PAF, it would be important to determine if, in intact cells, the addition of signals that stimulate acetyltransferase activity and the generation of PAF enhance the phosphorylation state of the enzyme.

As mentioned above, the addition of the ionophore A23187 to a variety of intact cells also result in the activation of 1-O-alkyl-2-lyso-sn-glycero-3-phosphocholine acetyltransferase, which has been interpreted as suggestive of a role for Ca^{2+} in the stimulation process. The addition of Ca^{2+} to splenic microsomes also results in stimulation of the acetyltransferase, this effect being reversed by the addition of EGTA.[50] The effect of Ca^{2+} was dose dependent, with a maximum of 0.1 to 10 μM Ca^{2+}. The addition of Mg^{2+}/ATP had no further effect on the acetylation reaction, and Ca^{2+} stimulated acetyltransferase in the absence of endogenous calmodulin.[50] These results are indicative of a role of Ca^{2+} in the activation process of acetyltransferase which seems to be independent of calmodulin and protein phosphorylation. In addition, Ca^{2+} appears to regulate PAF production at the level of phospholipase A_2-mediated production of lyso-PAF as Ca^{2+} is required to stimulate the hydrolysis of 1-O-alkyl-2-acyl-sn-glycero-3-phosphocholine.[89]

PAF is degraded by PAF acetylhydrolase, an enzyme which is highly specific for phospholipids with short acyl groups at the sn-2 position, and probably by the action of specific phospholipases C and D.[2,57,62] PAF acetylhydrolase has been shown to increase during macrophage differentiation in the absence of changes in the activities of the synthetic enzymes.[58] This effect was accompanied by a marked decrease in the accumulation of PAF in stimulated cells, suggesting that PAF acetylhydrolase can also regulate the generation of PAF.[58]

III. BIOLOGICAL ACTIONS OF PLATELET-ACTIVATING FACTOR

As mentioned in Section I, PAF stimulates many different target cells and, in each case, induces a set of specific biological responses. Of particular interest is the question of whether the activation of target cells by PAF is a receptor-mediated event. Irrefutable demonstration of the presence of receptors for lipid molecules is difficult. However, the results of a variety of experiments support the concept that some, if not all, actions of PAF are receptor mediated. Among these results is the finding that PAF is biologically active at very low concentrations (about 10^{-10} M) and the identification of monoclonal antibodies against platelets which specifically block PAF-induced stimulation.[63] Moreover, a variety of reports indicate the existence of specific PAF binding to isolated cells with high affinity and saturability.[64-66] Scatchard analysis of the binding experiments in platelets has shown two binding sites: a high-affinity (K_D, 37 nM), low-capacity binding site (about 1400 molecules bound per cell) and a second site with almost infinite capacity which might reflect nonspecific binding to PAF to the membranes.[64] Similar results have been obtained in a variety of studies using neutrophils and platelets from different species.[2,64-66] In guinea pig eosinophils, peroxidase release and intracellular Ca^{2+} mobilization occurred at much lower concentrations (1.3 nM and 11.5 nM) than superoxide anion generation (31.7 μM), which has been interpreted as suggestive of the existence of two different PAF receptors or a single receptor with two different affinity states.[90] The binding of PAF to isolated rabbit platelet membranes has been found to be inhibited by monovalent cations like Na^+ and Li^+, and to be stimulated by divalent cations like Mg^{2+} and Ca^{2+}.[66] A role for G proteins in PAF binding to membranes has also been proposed. Pertussis toxin, which is known to ADP-ribosylate the inhibitory guanyl nucleotide regulatory subunit of G proteins (see Chapter 7), has been shown to inhibit a variety of PAF-mediated effects in human neutrophils.[67] Moreover, a PAF-stimulated GTPase activity has been observed in rabbit platelet membranes,[66] which is inhibited by incubation with phorbol esters.[91] Phorbol esters have also been found to inhibit binding of PAF and the biological responses of neutrophils when triggered with this molecule.[92] Whether phorbol esters inhibit the binding of PAF through activation of protein kinase C or by perturbing the membrane directly is not known. Although these results strongly indicate that some of the biological effects of PAF are receptor mediated, purification of specific receptors remains to be accomplished. Recently, a photoreactive, biologically active, radioiodinated derivative of PAF was synthesized and used as a photoaffinity probe to study PAF binding sites in rabbit platelet membranes.[93] A protein with an M_r of 52 kDa was detected by SDS-polyacrylamide gel electrophoresis, and the labeling of this protein was inhibited by the presence of unlabeled PAF at nanomolar concentrations.[93]

Also favoring the existence of PAF receptors is the development of relatively specific PAF antagonists. Some of these molecules are PAF analogs with alterations in the substituents on the glycerol backbone. The best-characterized antagonists of PAF are CV 3988, a PAF analog whose structure is *rac*-3-(*N*-n-octadecyl-carbamoyloxy)-2-methoxypropyl-2-thiazolioethyl phosphate (Figure 4), BN 52021, which is extracted from *Ginkgo biloba* leaves, and the extract of the chinese plant Kadsurenone. These, and other compounds, have been reported to have agonist properties, desensitizing activity, or inhibitory PAF activity.[68-74] These are useful tools with which to investigate the pathophysiological role of PAF, and may also be helpful in the development of new antiinflammatory and antiallergic drugs.

Another type of molecule structurally related to PAF is 1-*O*-octadecyl-2-*O*-methyl-*rac*-glycero-3-phosphocholine and the sulphur-containing analog 1-*O*-hexadecyl-2-*O*-ethyl-*rac*-glycero-3-phosphocholine (Figure 4). These alkyl lysophospholipids, which have antineoplasic properties, do not seem to have a direct effect on DNA synthesis or function, and are nonmutagenic, inhibit the effects of phorbol diester tumor promoters and are effective inhibitors of protein kinase C.[75,76]

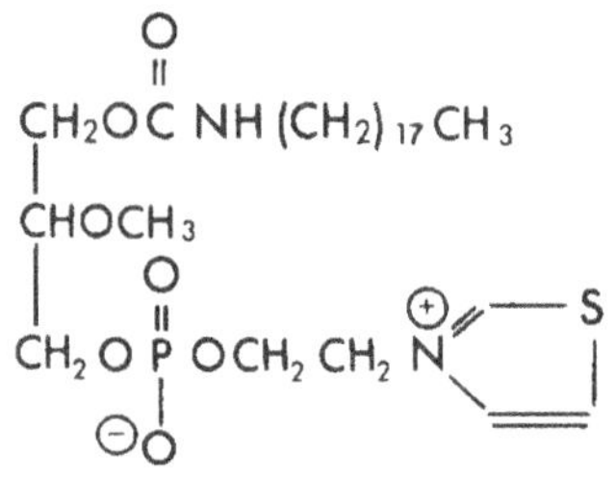

$$CH_2OC \overset{\overset{\displaystyle O}{\|}}{} NH\,(CH_2)_{17}CH_3$$

$$CHOCH_3$$

$$CH_2\,O\,\overset{\overset{\displaystyle O}{\|}}{\underset{\underset{\displaystyle \ominus O}{|}}{P}}\,OCH_2\,CH_2\,\overset{\oplus}{N}$$

CV3988

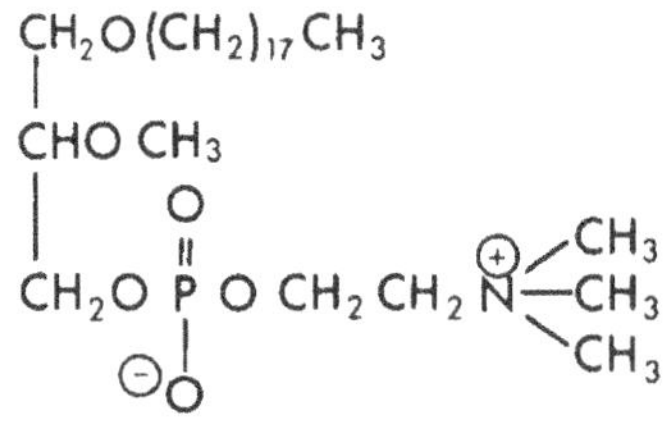

$$CH_2O\,(CH_2)_{17}CH_3$$

$$CHO\,CH_3$$

$$CH_2O\,\overset{\overset{\displaystyle O}{\|}}{\underset{\underset{\displaystyle \ominus O}{|}}{P}}\,O\,CH_2\,CH_2\,\overset{\oplus}{N}\!\!\begin{array}{l}-CH_3\\-CH_3\\-CH_3\end{array}$$

ET − 18 − OCH₃

$$CH_2\,SO_2\,(CH_2)_{15}CH_3$$

$$CHO\,CH_2\,CH_3$$

$$CH_2O\,\overset{\overset{\displaystyle O}{\|}}{\underset{\underset{\displaystyle \ominus O}{|}}{P}}\,O\,CH_2\,CH_2\,\overset{\oplus}{N}\!\!\begin{array}{l}-CH_3\\-CH_3\\-CH_3\end{array}$$

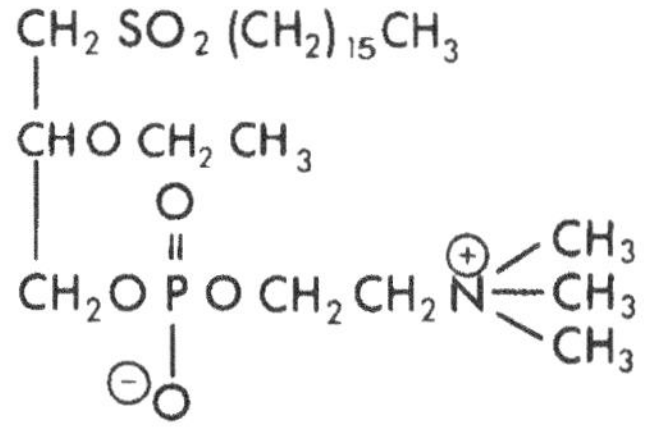

ET − 16S − OCH₂ − CH₃

FIGURE 4.   Structure of several PAF analogs: *rac*-3-(*N*-n-octadecyl-carbamoyloxy)-2-methoxypropyl-2-thiazolioethyl-phosphate, CV 3988; 1-*O*-octadecyl-2-*O*-methyl-*rac*-glycero-3-phosphocholine, ET-18-OCH₃; and 1-*O*-hexadecyl-2-*O*-ethyl-*rac*-glycero-3-phosphocholine, ET-16S-OCH₂-CH₃.

As mentioned in Section I, PAF is a potent lipid mediator of a variety of cellular processes, including platelet and polymorphonuclear neutrophil aggregation, as well as the production of thromboxane B and superoxide ions. PAF is also a potent hypotensive agent, and it has been shown to induce smooth muscle contraction and bronchoconstriction.[2,3] In addition, PAF is biologically active in many diverse biological systems. Thus, PAF (2 × 10⁻¹⁰ *M*) rapidly stimulates glycogenolysis in the intact perfused rat liver,[77] the effect being time, dose, and Ca²⁺ dependent.[78] However, the addition of PAF (10⁻⁹ *M*) to isolated rat hepatocytes, which were responsive to epinephrine, had no effect on glucose release.[79] This raises the intriguing question of how PAF stimulates glucose release in the perfused rat liver. Recently, it has been proposed that the interaction of PAF with Kupffer cells may result in the hemodynamic and metabolic responses in the liver.[95] Infusion of rat liver with heat-aggregated IgG stimulated both the production of PAF and the release of glucose, suggesting that the effect of PAF on glycogenolysis might be of physiological interest.[80] PAF has also been reported to have neuroregulatory and neuropathological actions. Thus, PAF increased the intracellular levels of Ca^{2+} in cells of the hybrid cholinergic clone NG108-15 and the adrenergic clone PC12 by stimulating the uptake of extracellular Ca^{2+}.[87] The effect of PAF on the uptake of Ca^{2+} by NG108-15 or PC12 cells was blocked by the PAF antagonist CV-3988, but not by other molecules such as triazolobenzodiazepines, triazolam, alprazolam, and brotizolam.[87] Moreover, PAF has been reported to stimulate the secretion of hypothalamic corticotropin-releasing hormone by retinal cells.[88]

Of particular interest is the question of how PAF initiates this variety of biological responses in target cells. PAF has been shown to induce in isolated rat hepatocytes a rapid decrease in the labeling of phosphatidylinositol-4,5-bisphosphate and phosphatidylinositol-4-phosphate with little or no effect on phosphatidylinositol turnover,[79] and to stimulate protein kinase C when added to human neutrophils.[81,82] Similarly, PAF has been shown to have an effect on exocytosis and a dose-dependent effect on Ca^{2+} uptake and phosphoinositide turnover when added to pancreatic lobules from the exocrine pancreas or the parotid glands of guinea pig.[83] PAF has been found to elicit an increase in the cytoplasmic levels of Ca^{2+}, as monitored by the fluorescent indicator Fura 2, and to stimulate arachidonic acid release in human polymorphonuclear neutrophils and mouse macrophages.[94,96] Increases in inositol phosphate production in response to PAF have been reported in a variety of cell types, including Kupffer cells,[95,97] rabbit and human platelets,[98,99] human endothelial and vascular smooth muscle,[100-101] and murine peritoneal macrophages.[102,103] Diacylglycerol accumulation has also been detected in response to PAF.[102,103] However, in murine peritoneal macrophages, down-regulation of protein kinase C by prolonged pretreatment with phorbol esters or inhibition of the enzyme with sphingosine-inhibited, PAF-dependent generation of diacylglycerol had no effect on the formation of inositol phosphates.[103] These results indicate that PAF stimulates the generation of diacylglycerol from sources other than the phosphatidylinositols, and that this step requires the activation of a protein kinase C.[103] Treatment of platelets or macrophages with PAF caused a time- and dose-dependent phosphorylation of several proteins,[98,102,104] which could be blocked by pretreatment of the cells with a variety of PAF antagonists.[104] Finally, PAF has also been reported to inhibit basal, prostaglandin E_1- and fluoride-stimulated adenylate cyclase,[84] and conditions that increase cAMP levels have been found to inhibit PAF generation.[85,86]

REFERENCES

1. **Snyder, F., Lee, T. C., and Wykle, R. L.,** Ether-linked glycerolipids and their bioactive species: enzymes and metabolic regulation, in *The Enzymes of Biological Membranes*, Vol. 2, Martonosi, A. N., Ed., Plenum Press, New York, 1985, 1.
2. **Hannahan, D. J.,** Platelet activating factor: a biologically active phosphoglyceride, *Annu. Rev. Biochem.,* 55, 483, 1986.
3. **Winslow, C. M. and Lee, M. L.,** *New Horizons in Platelet Activating Factor Research,* John Wiley & Sons, New York, 1987.
4. **Henson, P. M.,** Release of vasoactive amines from rabbit platelets induced by sensitized mononuclear leucocytes and antigen, *J. Exp. Med.,* 131, 287, 1970.
5. **Henson, P. M.,** The immunologic release of constituents from neutrophil leucocytes. II. Mechanisms of release during phagocytosis, and adherence to nonphagocytosable surfaces, *J. Immunol.,* 107, 1547, 1971.
6. **Benveniste, J., Henson, P. M., and Cochrane, C. G.,** Leukocyte-dependent histamine release from rabbit platelets, *J. Exp. Med.,* 136, 1356, 1972.

7. **Pinckard, R. N., Farr, R. S., and Hanahan, D. J.,** Physicochemical and functional identity of rabbit platelet activating factor (PAF) released *in vivo* during IgE anaphylaxis with PAF released *in vitro* from igE sensitized basophils, *J. Immunol.,* 123, 1847, 1979.
8. **Benveniste, J., Tencé, M., Varenne, P., Bidault, J., Boullet, C., and Polonsky, J.,** Semisynthèse et structure proposée du facteur activant les plaquettes (PAF): PAF-acéther, un alkyl éther analogue de la lysophosphatidylcholine, *C. R. Acad. Sci. Paris,* 289D, 1037, 1979.
9. **Blank, M. L., Snyder, F., Byers, L. W., Brooks, B., and Muirhead, E. E.,** Antihypertensive activity of an alkyl ether analog of phosphatidylcholine, *Biochem. Biophys. Res. Commun.,* 90, 1194, 1979.
10. **Hanahan, D. J., Demopoulos, C. A., Liehr, J., and Pinckard, R. N.,** Identification of platelet activating factor isolated from rabbit basophils as acetyl glyceryl ether phosphorylcholine, *J. Biol. Chem.,* 255, 5514, 1980.
11. **Mueller, H. W., O'Flaherty, J. T., and Wykle, R. L.,** The molecular species distribution of platelet-activating factor synthesized by rabbit and human neutrophils, *J. Biol. Chem.,* 259, 14554, 1984.
12. **Pinckard, R. N., Jackson, E. M., Hoppens, C., Weintraub, S. T., Ludwig, J. C., McManus, L. M., and Mott, G. E.,** Molecular heterogeneity of platelet-activating factor produced by stimulated human polymorphonuclear leucocytes, *Biochem. Biophys. Res. Commun.,* 122, 325, 1984.
13. **Weintraub, S. T., Ludwig, J. C., Mott, G. E., McManus, L. M., Lear, C., and Pinckard, R. N.,** Fast atom bombardment-mass spectrometric identification of molecular species of platelet-activating factor produced by stimulated human polymorphonuclear leucocytes, *Biochem. Biophys. Res. Commun.,* 129, 868, 1985.
14. **Ludwig, J. C. and Pinckard, R. N.,** Diversity in the chemical structure of neutrophil-derived platelet activating factors, in *New Horizons in Platelet Activating Factor Research,* Winslow, C. M. and Lee, M. L., Eds., John Wiley & Sons, New York, 1987, 59.
15. **Clay, K. L., Murphy, R. C., Andres, J. L., Lynch, J., and Henson, P. M.,** Structure elucidation of platelet activating factor derived from human neutrophils, *Biochem. Biophys. Res. Commun.,* 121, 815, 1984.
16. **Oda, M., Satouchi, K., Yasunaga, K., and Saito, K.,** Molecular species of platelet activating factor generated by human neutrophils challenged with ionophore A23187, *J. Immunol.,* 134, 1090, 1985.
17. **Mueller, H. W., O'Flaherty, J. T., Greene, D. G., Samuel, M. P., and Wykle, R. L.,** 1-O-alkyl-linked glycerophospholipids of human neutrophils: distribution of arachidonate and other acyl residues in the ether-linked and diacyl species, *J. Lipid Res.,* 25, 383, 1984.
18. **Satouchi, K., Oda, M., Yasunaga, K., and Saito, K.,** Evidence for the production of 1-acyl-2-acetyl-*sn*-glyceryl-3-phosphorylcholine concomitantly with platelet activating factor, *Biochem. Biophys. Res. Commun.,* 128, 1409, 1985.
19. **Debuch, H.,** The history of ether-linked lipids through 1960, in *Ether Lipids: Chemistry and Biology,* Snyder, F., Ed., Academic Press, New York, 1972, 1.
20. **Synder, F., Malone, B., and Wykle, R. L.,** The biosynthesis of alkyl ether bonds in lipids by a cell-free system, *Biochem. Biophys. Res. Commun.,* 34, 40, 1969.
21. **Snyder, F., Wykle, R. L., and Malone, B.,** A new metabolic pathway: Biosynthesis of alkyl ether bonds from glyceraldehyde-3-phosphate and fatty alcohols by microsomal enzymes, *Biochem. Biophys. Res. Commun.,* 34, 315, 1969.
22. **Hajra, A. K.,** Acyl dihydroxyacetone phosphate: Precursor of alkyl ethers, *Biochem. Biophys. Res. Commun.,* 39, 1037, 1970.
23. **Wykle, R. L., Piantadosi, C., and Snyder, F.,** The role of acyldihydroxyacetone phosphate, NADH, and NADPH in the biosynthesis of O-alkyl glycerolipids by microsomal enzymes of Ehrlich ascites tumour, *J. Biol. Chem.,* 247, 5442, 1972.
24. **Hajra, A. K.,** Biosynthesis of alkyl-ether containing lipid from dihydroxyacetone phosphate, *Biochem. Biophys. Res. Commun.,* 37, 486, 1969.
25. **Kennedy, E. P. and Weiss, S. B.,** The function of cytidine coenzymes in the biosynthesis of phospholipids, *J. Biol. Chem.,* 222, 193, 1956.
26. **Snyder, F., Blank, M. L., and Malone, B.,** Requirement of cytidine derivatives in the biosynthesis of O-alkyl phospholipids, *J. Biol. Chem.,* 245, 4016, 1970.
27. **Radominska-Pyrek, A. and Horrocks, L. A.,** Enzymic synthesis of 1-alkyl-2-acyl-*sn*-glycerol-3-phosphorylethanolamines by the CDP-ethanolamine:1-radyl-2-acyl-*sn*-glycerol ethanolaminephosphotransferase from microsomal fraction of rat brain, *J. Lipid Res.,* 13, 580, 1972.
28. **Roberti, R., Binaglia, L., Francescangeli, E., Goracci, G., and Porcelatti, G.,** Enzymatic synthesis of 1-alkyl-2-acyl-*sn*-glycero-3-phosphorylethanolamine through ethanolaminephosphotransferase activity in the neuronal and glial cells of rabbit *in vitro, Lipids,* 10, 121, 1975.
29. **Wykle, R. L., Malone, B., and Synder, F.,** Enzymatic synthesis of 1-alkyl-2-acetyl-*sn*-glycero-3-phosphocholine, a hypotensive and platelet aggregating lipid, *J. Biol. Chem.,* 255, 10256, 1980.
30. **Renooij, W. and Snyder, F.,** Biosynthesis of 1-alkyl-2-acetyl-*sn*-glycero-3-phosphocholine (platelet activating factor and a hypotensive lipid) by choline phosphotransferase in various rat tissues, *Biochim. Biophys. Acta,* 663, 545, 1981.

31. **Hajra, A. K.**, Biosynthesis of O-alkylglycerol ether lipids, in *Ether Lipids, Biochemical and Biomedical Aspects,* Mangold, H. K. and Paltauf, F., Eds., Academic Press, New York, 1985, 85.

32. **Snyder, F., Lee, T. C., and Wykle, R. L.**, Ether-linked glycerolipids and their bioactive species: enzymes and metabolic regulation, in *The Enzymes of Biological Membranes,* Vol. 2, Martonosi, A. N., Ed., Plenum Press, New York, 1985, 1.

33. **Malone, B., Lee, T. C., Snyder, F.**, De *novo* biosynthesis of alkylacetylglycerols, a precursor of platelet activating factor (PAF), *Fed. Proc.,* 45, 1529, 1986.

34. **Lee, T. C., Malone, B., and Snyder, F.**, A new de *novo* pathway for the formation of 1-alkyl-2-acetyl-*sn*-glycerols, a precursor of platelet activating factor. Biochemical characterization of 1-alkyl-2-lyso-*sn*-glycero-3-P:acetyl-CoA acetyltransferase, *J. Biol. Chem.,* 261, 5373, 1986.

35. **Blank, M. L., Lee, T. C., Cress, E. A., Malone, B., Fitzgerald, V., and Snyder, F.**, Conversion of 1-alkyl-2-acetyl-*sn*-glycerols to platelet-activating factor and related phospholipids by rabbit platelets, *Biochem. Biophys. Res. Commun.,* 124, 156, 1984.

36. **Record, M. and Snyder, F.**, Biosynthesis of platelet-activating factor (PAF) via alternate pathways: subcellular distribution of products in HL-60 cells, *Fed. Proc.,* 45, 1529, 1986.

37. **Fernandez-Gallardo, S., Gijon, M. A., Garcia, M. C., Cano, E., and Sanchez Crespo, M.**, Biosynthesis of platelet-activating factor in glandular gastric mucosa, *Biochem. J.,* 254, 707, 1988.

38. **Alonso, F., Garcia Gil, M., Sanchez Crespo, M., and Mato, J. M.**, Activation of 1-alkyl-2-lyso-glycero-3-phosphocholine: acetyl-CoA transferase during phagocytosis in human polymorphonuclear leucocytes, *J. Biol. Chem.,* 257, 3376, 1982.

39. **Mencia-Huerta, J. M., Roubin, R., Morgat, J. L., and Benveniste, J.**, Biosynthesis of platelet activating factor (PAF-acether). III. Formation of PAF-acether from synthetic substrates by stimulated murine macrophages, *J. Immunol.,* 129, 804, 1982.

40. **Lee, T. C., Malone, B., Wassermann, S. I., Fitzgerald, V., and Snyder, F.**, Activities of enzymes that metabolize platelet-activating factor (1-alkyl-2-acetyl-*sn*-glycero-3-phosphocholine) in neutrophils and eosinophils from humans and the effect of a calcium ionophore, *Biochem. Biophys. Res. Commun.,* 105, 1303, 1982.

41. **Albert, D. H. and Snyder, F.**, Biosynthesis of 1-alkyl-2-acetyl-*sn*-glycero-3-phosphocholine (platelet activating factor) from 1-alkyl-2-acyl-*sn*-glycero-3-phosphocholine by rat alveolar macrophages: ionophore stimulation, *J. Biol. Chem.,* 258, 97, 1982.

42. **Roubin, R., Mencia-Huerta, J. M., Landes, A., and Benveniste, J.**, Biosynthesis of platelet-activating factor (PAF-acether). IV. Impairment of acetyltransferase activity in thioglycollate-elicited mouse macrophages, *J. Immunol.,* 129, 809, 1982.

43. **Snyder, F., Blank, M. L., Johnson, D., Lee, T. C., Malone, B., Robinson, M., and Woodard, D.**, Alkylacetylglycerols versus lyso-PAF as precursors in PAF biosynthesis and the role of arachidonic acid, *Pharmacol. Res. Commun.,* 17, 657, 1986.

44. **Snyder, F.**, The significance of dual pathways for the biosynthesis of platelet activating factor: 1-alkyl-2-lyso-*sn*-glycero-3-phosphate as a branchpoint, in *New Horizons in Platelet Activating Factor Research,* Winslow, C. M. and Lee, M. L., Eds., John Wiley & Sons, New York, 1987, 13.

45. **Lenihan, D. J. and Lee, T. C.**, Regulation of platelet-activating factor synthesis: modulation of 1-O-alkyl-2-lyso-*sn*-glycero-3-phosphocholine:acetyl-CoA acetyltransferase by phosphorylation and dephosphorylation in rat splenic microsomes, *Biochem. Biophys. Res. Commun.,* 120, 834, 1984.

46. **Nieto, M. L., Velasco, S., and Sanchez-Crespo, M.**, Modulation of acetyl-CoA:1-alkyl-2-lyso-*sn*-glycero-3-phosphocholine (lyso-PAF) acetyltransferase in human polymorphonuclears, *J. Biol. Chem.,* 263, 4607, 1988.

47. **Gomez-Cambronero, J., Velasco, S., Mato, J. M., and Sanchez-Crespo, M.**, Modulation of acetyl-transferase activity in rat splenic microsomes. II. The role of catalytic subunit of cyclic AMP-dependent protein kinase, *Biochim. Biophys. Acta,* 845, 516, 1985.

48. **Gomez-Cambronero, J., Velasco, S., Sanchez-Crespo, M., Vivanco, F., and Mato, J. M.**, Partial purification and characterization of 1-O-alkyl-2-lyso-*sn*-glycero-3-phosphocholine:acetyl-CoA acetyltransferase from rat spleen, *Biochem. J.,* 237, 439, 1986.

49. **Gomez-Cambronero, J., Mato, J. M., Vivanco, F., and Sanchez-Crespo, M.**, Phosphorylation of partially purified 1-O-alkyl-2-lyso-*sn*-glycero-3-phosphocholine:acetyl-CoA acetyltransferase from rat spleen, *Biochem. J.,* 245, 893, 1987.

50. **Gomez-Cambronero, J., Nieto, M., Mato, J. M., and Sanchez-Crespo, M.**, Modulation of acetyltransferase activity in rat splenic microsomes. I. The role of calcium ions, *Biochim. Biophys. Acta,* 845, 511, 1985.

51. **Kramer, R. M., Patton, G. M., Pritzker, C. R., and Deykin, D.**, Metabolism of platelet activating factor in human platelets. Transacylase-mediated synthesis of 1-O-alkyl-2-lyso-*sn*-glycero-3-phosphocholine:acetyl CoA acetyltransferase in rat spleen microsomes, *J. Biol. Chem.,* 260, 10952, 1984.

52. **Robinson, M., Blank, M. L., and Snyder, F.**, Acylation of lysophospholipids by rabbit alveolar macrophages. Specificities of CoA-dependent and CoA-independent reactions, *J. Biol. Chem.,* 260, 7889, 1985.

53. **Sugiura, T. and Waku, K.,** CoA-independent transfer of arachidonic acid from 1,2-diacyl-*sn*-glycero-3-phosphocholine to 1-O-alkyl-*sn*-glycero-3-phosphocholine (lyso platelet activating factor) by macrophage microsomes, *Biochem. Biophys. Res. Commun.,* 127, 384, 1985.

54. **Masuzawa, Y., Okano, S., Nagakawa, Y., Ojima, A., and Waku, K.,** Selective acylation of alkyllysophospholipids by docohexaenoic acid in Ehrlich ascites cells, *Biochim. Biophys. Acta,* 876, 80, 1986.

55. **Ojima, A., Nakagawa, Y., Sugiura, T., Masuzawa, Y., and Waku, K.,** Selective transacylation of 1-O-alkylglycerophosphoethanolamine by docohexaenoate and arachidonate in rat brain microsomes, *J. Neurochem.,* 48, 1403, 1987.

56. **Ninio, E., Joly, F., Hieblot, C., Bessou, G., Mencia-Huerta, J. M., and Benveniste, J.,** Biosynthesis of PAF-acether. IX. Role for a phosphorylation-dependent activation of acetyltransferase in antigen-stimulated mouse mast cells, *J. Immunol.,* 139, 154, 1987.

57. **Wilcox, R. W., Wykle, R. L., Schmitt, J. D., and Daniel, L. W.,** The degradation of platelet-activating factor and related lipids: susceptibility to phospholipase C and D, *Lipids,* 22, 800, 1987.

58. **Elstad, M. R., Stafforini, D. M., McIntyre, T. M., Prescott, S. M., and Zimmerman, G. A.,** Platelet-activating factor acetylhydrolase increases during macrophage differentiation, *J. Biol. Chem.,* 264, 8467, 1989.

59. **Woodard, D. S., Lee, T. C., and Snyder, F.,** The final step in the *de novo* biosynthesis of platelet activating factor. Properties of a unique CDP-choline:1-alkyl-2-acetyl-*sn*-glycerol choline-phosphotetransferase in microsomes from the renal inner medula of rats, *J. Biol. Chem.,* 262, 2520, 1987.

60. **Lee, T. C., Malone, B., Snyder, F.,** Formation of 1-alkyl-2-acetyl-*sn*-glycerols via *de novo* biosynthetic pathway for platelet activating factor. Characterization of 1-alkyl-2-acetyl-*sn*-glycero-3-phosphate phosphohydrolase in rat spleen, *J. Biol. Chem.,* 263, 1755, 1988.

61. **Blank, M. L., Lee, J. Y., Cress, E. A., and Snyder, F.,** Stimulation of the *de novo* pathway for the biosynthesis of platelet activating factor (PAF) via cytidylyltransferase activation in cells with minimal endogenous PAF production, *J. Biol. Chem.,* 263, 5656, 1988.

62. **Farr, R. S., Wardlow, M. L., Cox, C. P., Meng, K. E., and Greene, D. R.,** Human-serum acid-labile factor is acylhydrolase that inactivates platelet activating factor, *Fed. Proc.,* 42, 3121, 1983.

63. **Lynch, J. M., Worthen, G. S., and Henson, P. M.,** Platelet activating factor, in *The Development of Anti-Asthmatic Drugs,* Buckle, E. and Smith, H., Eds., Butterworths, London, 1982, 73.

64. **Valone, F. H.,** Isolation of a platelet membrane protein which binds the platelet activating factor 1-O-hexadecyl-2-acetyl-*sn*-glycero-3-phosphocholine, *Immunology,* 52, 169, 1984.

65. **Kloprogge, E. and Akkerman, J. W. N.,** Binding kinetics of PAF-acether (1-O-alkyl-2-acetyl-*sn*-glycero-3-phosphocholine) to intact human platelets, *Biochem. J.,* 223, 901, 1984.

66. **Hwang, S., Lam, M., and Pong, S.,** Ionic and GTP regulation of binding of platelet activating factor to receptors and platelet activating factor-induced activation of GTPase in rabbit platelet membranes, *J. Biol. Chem.,* 261, 532, 1986.

67. **Lad, P. M., Olson, C. V., and Grewal, I. S.,** Platelet activating factor mediated effects on human neutrophil function are inhibited by pertussis toxin, *Biochem. Biophys. Res. Commun.,* 129, 632, 1985.

68. **Terashita, Z., Tsushima, S., Yoshioka, Y., Nomura, H., Inada, Y., Nishikawa, K.,** CV-3988 — a specific antagonist of platelet-activating factor, *Life Sci.,* 32, 1975, 1983.

69. **Venuti, M. C.,** Platelet-activating factor: multifaceted biochemical and physiological mediator, *Annu. Rep. Med. Chem.,* 20, 193, 1985.

70. **Shen, T. Y., Hwang, S. B., Chang, N. M., Doebber, T. W., Lam, M. H. T., Wu, M. S., Wang, X., Han, G. Q., and Li, R. Z.,** Characterization of a platelet-activating factor antagonist isolated from Haifenteng (Piper futokadsura): specific inhibition of *in vitro* and *in vivo* platelet-activating factor-induced effects, *Proc. Natl. Acad. Sci. U.S.A.,* 82, 672, 1985.

71. **Braquet, P.,** Treatment and Prevention of PAF-Acether-Induced Sickness by a New Series of Highly Specific Inhibitors, *British Patent* 8, 418, 1984.

72. **Nunez, P. and Benveniste, J.,** Specific inhibition of PAF-acether-induced platelet activation by BN 52021 and comparison with the PAF-acether inhibitors kadsurenone and CV 3988, *Eur. J. Pharmacol.,* 123, 197, 1986.

73. **Joly, F., Bessou, G., Benveniste, J., and Ninio, E.,** Ketotifen inhibits PAF-acether biosynthesis and β-hexosaminidase release in mouse mast cells stimulated with antigen, *Eur. J. Pharmacol.,* 144, 133, 1987.

74. **Barzaghi, G. and Mong, S.,** Platelet activating factor-amidophosphonate (PAF-AP), a partial agonist inhibited platelet activating factor-induced calcium mobilization in human monocytic leukemic U-937 cell, *Life Sci.,* 44, 361, 1989.

75. **Morris-Natschke, S., Surles, J. R., Daniel, L. W., Berens, M. E., Modest, E. J., and Piantadosi, C.,** Synthesis of sulphur analogues of alkyl lysophospholipid and neoplastic cell growth inhibitory properties, *J. Med. Chem.,* 29, 2114, 1986.

76. **Daniel, L. W., Etkin, L. A., Morrison, B. T., Parker, J., Morris-Natschke, S., Surles, J., and Piantadosi, C.,** Ether lipids inhibit the effects of phorbol diester tumour promoters, *Lipids,* 22, 851, 1987.

77. Shukla, S. D., Buxton, D. B., Olson, M. S., and Hanahan, D. J., Acetyl glyceryl ether phosphoryl-choline. A potent activator of hepatic phosphoinositide metabolism and glycogenolysis, *J. Biol. Chem.*, 258, 10212, 1983.

78. Buxton, D. B., Shukla, S. D., Hanahan, D. J., and Olson, M. S., Stimulation of hepatic glycogenolysis by acetyl glyceryl ether phosphoryl choline, *J. Biol. Chem.*, 259, 1468, 1984.

79. Fisher, R. A., Shukla, S. D., DeBuysere, M., Hanahan, D. J., and Olson, M. S., The effect of acetyl glyceryl ether phosphorylcholine on glycogenolysis and phosphatidylinositol-4,5-bisphosphate in rat hepatocytes, *J. Biol. Chem.*, 259, 8685, 1984.

80. Buxton, D. B., Hanahan, D. J., and Olson, M. S., Stimulation of glycogenolysis and platelet activating factor production by heat-aggregated immunoglobulin G in the perfused rat liver, *J. Biol. Chem.*, 259, 13568, 1984.

81. Gay, J. C. and Stitt, E. S., Platelet activating factor induces protein kinase activity in the particulate fraction of human neutrophils, *Blood*, 71, 159, 1988.

82. O'Flaherty, J. and Nishihira, J., Arachidonate metabolites, platelet activating factor, and the mobilization of protein kinase C in human polymorphonuclear neutrophils, *J. Immunol.*, 138, 1889, 1987.

83. Söling, H. D., Eibl, H., and Fest, W., Acetylcholine-like effects of 1-O-alkyl-2-acetyl-*sn*-glycero-3-phosphocholine (PAF) and its analogues in exocrine secretory glands, *Eur. J. Biochem.*, 144, 65, 1984.

84. Haslam, R. J. and Vanderwel, M., Inhibition of platelet adenylate cyclase by 1-O-alkyl-2-O-acetyl-*sn*-glyceryl-3-phosphorylcholine (platelet activating factor), *J. Biol. Chem.*, 257, 6879, 1982.

85. Bussolino, F. and Benveniste, J., Pharmacological modulation of platelet-activating factor (PAF) release from rabbit leucocytes. I. Role of cyclic AMP, *Immunology*, 40, 367, 1980.

86. Alonso, F., Sanchez-Crespo, M., and Mato, J. M., Modulatory role of cyclic AMP in the release of platelet-activating factor for human polymorphonuclear leucocytes, *Immunology*, 45, 493, 1982.

87. Kornecki, E. and Ehrlich, Y. H., Neuroregulatory and neuropathological actions of the ether-phospholipid platelet-activating factor, *Science*, 240, 1792, 1988.

88. Leyravaud, S., Bossant, M. J., Joly, F., Bessou, G., Benveniste, J., and Ninio, E., Biosynthesis of PAF-acether. X. Phorbol myristate acetate-induced PAF-acether biosynthesis and acetyltransferase activation in human neutrophils, *J. Immunol.*, 143, 245, 1989.

89. Whatley, R. E., Nelson, P., Zimmerman, G. A., Stevens, D. L., Parker, C. J., McIntyre, T. M., and Prescott, S. M., The regulation of platelet-activating factor production in endothelial cells. The role of calcium and protein kinase C, *J. Biol. Chem.*, 264, 6325, 1989.

90. Kroegel, C., Yukawa, T., Westwick, J., and Barnes, P. J., Evidence for two platelet activating factor receptors on eosinophils: dissociation between PAF-induced intracellular calcium mobilization, degranulation and superoxide anion generation in eosinophils, *Biochem. Biophys. Res. Commun.*, 162, 511, 1989.

91. Avdonin, P. V., Svitina-Ulitina, I. V., and Tkachnuk, V. A., Selective inactivation by endogenous protein kinase C of human platelet high-affinity GTPase coupled with PAF receptors, *J. Mol. Cell. Cardiol.*, s1, 139, 1989.

92. Yamazaki, M., Gomez-Cambroner, J., Durstin, M., Molski, T. F., Becker, E. L., and Sha'afi, S., Phorbol 12-myristate 13-acetate inhibits binding of leukotriene B₄ and platelet activating factor and the responses they induce in neutrophils: sites of action, *Proc. Natl. Acad. Sci. U.S.A.*, 86, 5791, 1989.

93. Chau, L. Y., Tsai, Y. M., and Gheng, J. R., Photoaffinity labelling of platelet activating factor binding sites in rabbit platelet membranes, *Biochem. Biophys. Res. Commun.*, 161, 1070, 1989.

94. Nakashima, S., Suganuma, A., Sato, M., Tohmatsu, T., and Nozawa, Y., Mechanism of arachidonic acid liberation in platelet-activating factor-stimulated human polymorphonuclear neutrophils, *J. Immunol.*, 143, 1295, 1989.

95. Fisher, R. A., Sharma, R. V., and Bhalla, R. C., Platelet-activating factor increases inositol phosphate production and cytosolic free Ca^{2+} concentrations in cultured rat Kupffer cells, *FEBS Lett.*, 251, 22, 1989.

96. Randriamampita, C. and Trautmann, A., Biphasic increase in intracellular calcium induced by platelet-activating factor in macrophages, *FEBS Lett.*, 249, 199, 1989.

97. Bankey, P. E., Billiar, T. R., Wang, W. Y., Carlson, A., Holman, R. T., and Cerra, F. B., Modulation of Kupffer cell membrane phospholipid function by n-3 polyunsaturated fatty acids, *J. Surg. Res.*, 46, 439, 1989.

98. Morrison, W. J., Dhar, A., and Shukla, S. D., Staurosporine potentiates platelet activating factor stimulated phospholipase C activity in rabbit platelets but does not block desensitization by platelet activating factor, *Life Sci.*, 45, 333, 1989.

99. Shukla, S. D., Morrison, W. J., and Klachko, D. M., Response to platelet-activating factor in human platelets stored and aged in plasma. Decrease in aggregation, phosphoinositide turnover, and receptor affinity, *Transfusion*, 29, 528, 1989.

100. Grigorian, G. V., Mirzapoyazova, T. Y., Resink, T. J., Danilov, S. M., and Tkachuk, V. A., Regulation of phosphoinositide turnover in endothelium from human pulmonary artery, aorta and umbilical vein. Antagonistic action of the beta-adrenoreceptor coupled adenylate cyclase system, *J. Mol. Cell. Cardiol.*, s1, 119, 1989.

101. **Shirinsky, V. P., Sobolevsky, A. V., Grigorian, G. V., Danilov, S. M., Tararak, E. M., and Tkachuk, V. A.**, Agonist-induced polyphosphoinositide breakdown in cultured human endothelial and vascular smooth muscle cells, *Health Psychol.,* s7, 61, 1988.
102. **Prpic, V., Uhing, R. J., Weiel, J. E., Jakoi, L., Gawdi, G., Herman, B., and Adams, D. O.**, Biochemical and functional responses stimulated by platelet-activating factor in murine peritoneal macrophages, *J. Cell. Biol.,* 107, 363, 1988.
103. **Uhing, R. J., Prpic, V., Hollenbach, P. W., and Adams, D. O.**, Involvement of protein kinase C in platelet-activating factor-stimulated diacylglycerol accumulation in murine peritoneal macrophages, *J. Biol. Chem.,* 264, 9224, 1989.
104. **Shukla, S. D., Morrison, W. J., and Dhar, A.**, Desensitization of platelet-activating factor-stimulated protein phosphorylation in platelets, *Mol. Pharmacol.,* 35, 409, 1989.

INDEX

Printed and bound by CPI Group (UK) Ltd, Croydon, CR0 4YY

22/10/2024

01777632-0012